TESTING LOUDSPEAKERS

D1580717

TESTING LOUDSPEAKERS

by Joseph D'Appolito, Ph.D.

Published by

KCK Media Corp.

FIRST EDITION
1998
Copyright © 1998 by Joseph D'Appolito

Published by

KCK Media Corp.

Library of Congress Card Catalog Number: 98-072820-L

ISBN: 1-882580-17-6

First Printing August 1998

This book is affectionately dedicated to the women in my life: my wife, Ligia, my daughters, Daniela and Nina, and my granddaughters, Emma and Isabella. Their love and support and especially their enthusiasm for this book helped me through the many long hours of writing and the frequent writer's blocks that are part of creating a work of this size.

Acknowledgment

Any success this book enjoys will be due in no small part to the unstinting efforts of Richard Campbell. Dick, a professor, a Fellow of the Audio Engineering Society, and head of his own consulting firm, read every chapter of this book in draft form. He caught several errors in the text and offered many valuable suggestions for improving the presentation. His timely response to each chapter submission substantially re-duced the time between the creation and publication of this work.

TABLE OF CONTENTS

TESTING LOUDSPEAKERS

CHAPTER 1

AN INTRODUCTION TO *TESTING LOUDSPEAKERS*

1.1 WHY WRITE A BOOK ON LOUDSPEAKER TESTING?

There is no lack of books and articles on loudspeaker design and construction. Sources range from "how-to" construction articles in hobbyist magazines to books such as *The Loudspeaker Design Cookbook (LDC)*[1] to engineering and scientific articles in professional and academic journals. Material on loudspeaker testing, however, is not as plentiful or as well organized; it is scattered in journal papers, textbooks, manufacturers' brochures, and test standards.

In journal and magazine articles, test results are often presented in support of the main thrust of the paper. In this case, test procedures are rarely spelled out in detail, and limits on testing accuracy are not discussed. To my knowledge, there is currently no well-organized tutorial reference on loudspeaker testing for the experienced hobbyist or technician. This book is intended to fill that gap.

1.2 WHAT MATERIAL SHOULD A BOOK ON LOUDSPEAKER TESTING COVER?

A loudspeaker is a complex electro-mechano-acoustical transducer. It takes electrical energy at its input and converts it into mechanical motion, which in turn generates an acoustic output. Understanding loudspeaker testing is therefore a multi-disciplinary venture requiring an appreciation of pertinent principles from electrical, mechanical, and acoustical engineering. A broad understanding of these principles is required to select appropriate tests and to properly interpret test results in the presence of confounding environmental effects.

I believe a book on loudspeaker testing must take more than a simple "how-to" approach. It must present in an interesting and intuitive way enough applicable theory—supported with specific examples—from electrical, mechanical, and acoustical engineering. This should enable you to select appropriate tests and conduct them in a manner that will minimize errors and maximize the amount of useful information obtained from them.

To accomplish this goal the book must be tutorial. I have spent a great deal of time explaining the background material needed to understand loudspeaker testing. Many examples are presented to make the concepts clear. All examples are taken from real tests. There are no made up academic exercises. Chapters typically contain 60–80 or more figures and tables describing real experiments or tests. In many cases you can follow the step-by-step procedures, seeing all of the intermediate results leading to the final result.

1.3 WHAT SHOULD YOU KNOW TO GET THE MOST OUT OF THIS BOOK?

I expect you to be familiar with the basic principles of loudspeaker design as set forth in *LDC*. This includes some familiarity with Thiele/Small parameters for characterizing drivers and their use in the design of low-frequency direct radiator loudspeaker configurations such as the closed-box, vented-box, bandpass, and transmission-line systems. You should have a general appreciation for the various types of crossover networks and their use in multi-way loudspeaker systems. At a very fundamental level you should also be familiar with terms such as voltage, current, impedance, inductance, capacitance, resistance, frequency, displacement, velocity, acceleration, damping, mass, and pressure.

Mathematics is unavoidable in loudspeaker testing. Most test results are quantitative. I have tried to keep equations to a minimum, but in many instances they are the best way to describe a principle or concept. You will not need calculus, but you will need some background in algebra. You should know what logarithms and exponentials are and be able to manipulate them in equations. You should also know how to use a calculator to take logs or raise numbers to a power.

Beyond these basic requirements the book is largely self-contained. All of the concepts and engineering principles that I believe you will need to understand loudspeaker testing are described and illustrated with examples in the text.

1.4 THE BOOK IN BRIEF

Testing Loudspeakers is divided into two broad areas that might humorously be labeled BC and AD. That is, before computers and after digital. Chapters 2–5 cover classical analog techniques for measuring impedance, Thiele/Small parameters, low-frequency enclosure alignments, and frequency response using relatively simple and widely available analog test equipment. Chapters 6 and 7 cover the use of PC-based electrical and acoustical data-acquisition and analysis systems in loudspeaker testing. A short summary of each chapter follows.

1.4.1 CHAPTER 2: DRIVER TESTING

Chapter 2 begins with a brief history of the evolution of low-frequency direct radiator loudspeaker system design starting with the invention of the direct radiator loudspeaker by Kellogg and Rice and continuing through to the loudspeaker synthesis techniques contained in the landmark papers by Thiele, Benson, and Small. The Thiele/Small (T/S) parameters that characterize low-frequency loudspeaker behavior are defined and a development of their meaning from physical principles is presented. Although these parameters characterize the full electrical, mechanical, and acoustical performance of a driver, they can all be determined from measurements of loudspeaker voice-coil impedance. That is, they can all be determined from the electrical input side of the transducer.

Because the voice-coil impedance curve plays such a major role in determining T/S parameters, a driver electromechanical model is developed from first principles and used to explain the shape of the impedance curve. This development introduces and explains terms such as cone mass, suspension compliance and damping, voice-coil inductance and resistance, magnetic flux density, and resonant frequency. This leads to a precise physical definition of Q. An entire section is also devoted to a detailed explanation of V_{AS}, the air spring equivalent of driver compliance.

Three techniques are described for measuring the driver impedance curve and thus determining the free-air T/S parameters. They are the voltage-divider method, and the constant-current and constant-voltage methods. All of these measurements can be made with relatively simple analog test equipment, which includes an audio oscillator, a wideband AC voltmeter, and a frequency counter. Step-by-step procedures are given for each method, complete with examples illustrating the results at each step.

Of the three approaches, the voltage-divider technique is the simplest to apply, but also in its uncompensated form, the least accurate. A simple procedure is described for correcting errors in the voltage-divider approach. The several examples then show that all three methods yield comparable results when properly applied. For each method, several suggestions are made for avoiding errors and maximizing the accuracy of the measurements.

1.4.2 CHAPTER 3: LOW-FREQUENCY SYSTEM ELECTRICAL IMPEDANCE TESTS

In this chapter I examine the impedance characteristics of drivers mounted in enclosures. The impedance curves for drivers in closed- and vented-boxes are used to determine parameters such as system resonant frequency or frequencies, enclosure damping and Q, system efficiency, and the effect of enclosure filling material.

The chapter begins with a theoretical discussion of the mass reactance loading on a driver cone that occurs when a driver is mounted on a large baffle or in an enclosure. This loading causes a downward shift in driver resonant frequency. The impedance curve for a driver mounted in a closed box is then presented together with procedures for measuring the mass reactance loading and the downward shift in driver resonant frequency, the driver/box system resonant frequency and compliance ratio, α, and the system electrical and mechanical Qs. The procedures are illustrated with a real example and the measured results compared against earlier theoretical predictions.

The effect of filling material on closed-box systems is next explained in terms of gas dynamics. Filling material such as fiberglass or Dacron is shown to increase the effective volume of the enclosure, increase mass reactance loading on the driver cone, increase system losses, and decrease box Q and system efficiency. A procedure is given for measuring these effects and illustrated with an example.

The impedance curve of single-tuned vented systems is then described and procedures given for measuring driver resonant frequency and mass

reactance loading in a vented box, the box tuning frequency, the driver/box compliance ratio, and driver and box Qs. Two examples are given. An interesting dual woofer example illustrates the increase in mass reactance loading caused by inter-driver mutual coupling.

The chapter continues with an examination of the impedance curve for a transmission-line enclosure with and without line filling material. The impedance curve with filling material clearly shows the increased damping and mass reactance loading of the driver cone.

Chapter 3 ends with a discussion of anomalous impedance behavior and what can be learned from it. Unexpected wiggles, glitches, or jumps in the impedance curve are shown to reflect the effects of cabinet wall vibrations, internal standing waves, and, in the case of vented systems, the presence of "organ pipe" resonances in the port tubes. Four examples of anomalous impedance behavior are given. The detective work required to isolate the source of anomalous impedance in each case is explained.

1.4.3 CHAPTER 4: ACOUSTICAL TESTING OF SINGLE DRIVERS

Chapter 4 examines the factors influencing the accuracy of frequency-response tests of individual drivers mounted in enclosures. After describing the frequency-response curve, this chapter discusses microphone types and shows which microphones are best for loudspeaker testing. Microphones are classified according to their transduction principle, the acoustic quantity they sense, and their polar response. Omnidirectional pressure microphones calibrated for free-field response are shown to be best for loudspeaker frequency-response measurements.

Ideally, the measured frequency response should include only the direct-field response of the driver or loudspeaker system including the effect of baffle geometry. Unfortunately, acoustic measurements are strongly influenced by their environment. Standing waves, late arrivals of acoustic energy from reflecting surfaces, and ambient noise can all corrupt the measurement. Much of the art of loudspeaker measurement deals with recognizing these errors and minimizing or eliminating them from the frequency-response data.

The remainder of the chapter examines the effect of the acoustic environment and baffle geometry on frequency-response measurements and details various test techniques and test signals that can be used to recognize and mitigate environmental effects. Many examples are provided to illustrate the material in this chapter.

Since most readers will not have access to an anechoic chamber, the emphasis in Chapter 4 is placed on obtaining accurate measurements in typical listening rooms—large rooms with reflecting surfaces. The development shows that there are two distinct environmental error sources operating in disjoint frequency ranges—standing waves and surface reflections. For typical rooms, standing waves below the 200–400Hz range are shown to produce large, widely separated dips and peaks in frequency response which cannot be averaged out. Above this frequency range standing waves and surface reflections tend to be more uniform in amplitude and more evenly spaced in frequency.

An equation is given for predicting standing-wave modes in a rectangular room and an example calculation presented. The dominant floor reflection and the comb-filter response it produces are also described in detail, and equations are given to predict the frequency of the dips and peaks this reflection causes. Frequency-response data, which illustrates the comb-filter response and verifies the predictive equation, is then presented.

In Chapter 4 I describe three test techniques and two test signals that greatly reduce the impact of standing waves and reflections on measurement accuracy. The test techniques are far-field, near-field, and ground plane measurements. Pink noise and warble tones are the test signals.

Near-field measurements, which eliminate the effect of low-frequency standing waves, are described. Ground-plane measurements are shown to reduce errors caused by surface reflections. Pink noise and warble tone test signals used with either ground-plane or far-field measurements are shown to average out the effect of higher-frequency reflections. Several examples are presented to illustrate and compare each approach. Procedures are also given for correctly merging near-field and far-field measurements to produce a full-range frequency response.

The effect of baffle geometry on frequency response is described in detail in Chapter 4. The two most important effects are edge diffraction and low-frequency diffraction loss, also called spreading loss. Diffraction-induced frequency-response variations that are part of the overall speaker response must be recognized and separat ed from response variations due to environmental effects. Simple formulas are given for predicting the frequencies or frequency ranges over which diffraction effects will be observed. Examples that illustrate diffraction effects and confirm the formula predictions are then presented.

Chapter 4 ends with two interesting examples. The first illustrates how the frequency-response curves of multiple drivers can be used to select a crossover frequency. The second example describes an alternate and usually more accurate technique for determining the tuning frequency of a vented box using a near-field measurement of the woofer acoustic output.

1.4.4 CHAPTER 5: ACOUSTICAL TESTING OF MULTIPLE-DRIVER SYSTEMS

The single most important measure of loudspeaker system quality is frequency response, both on-axis and off-axis. The discussion of acoustical testing begun in Chapter 4 is extended in Chapter 5 to encompass frequency-response measurements of multiple-driver systems. All of the environmental considerations covered in Chapter 4 apply to the testing of multiple-driver systems. There are two additional issues unique to multiple-driver systems—driver integration and multiple-driver floor bounce.

Because the acoustic centers of the individual drivers in multidriver systems are not coincident, the distances from each driver to the listening or measurement position are not the same. Proper microphone placement for accurate frequency-response measurement of multidriver systems poses a problem. As the test microphone position is varied, signals from the individual drivers will arrive with varying phase, which, in turn, will change the way these signals add together.

Additionally, if the microphone is placed too close to the system, the drivers may not integrate properly. If the microphone is placed too far from the system, the measurements will be contaminated by room reflections and ambient noise. To assess the magnitude of this problem, I present an example showing the changes in response of a typical three-way system encountered as microphone location is varied.

This example also shows that the changes experienced are not only a function of microphone position, but are also strongly influenced by the order of the crossover network used. Odd-order crossovers are shown to produce much larger response variations with microphone position than even-order crossovers. These results emphasize the need to carefully choose the system design axis and test microphone position.

The simple floor bounce model developed in Chapter 4 is extended in Chapter 5 to include the multiple-driver case. Here individual drivers are at different heights off the floor and not at the same height as the microphone. More general equations are developed for predicting floor bounce frequencies for each driver. In an example, floor bounce frequencies of the woofer and midrange drivers are staggered so that dips in one can be filled in with peaks from the other. The example also shows how to eliminate the first floor bounce of each driver with an appropriate selection of the crossover frequency.

Chapter 5 continues with several examples of frequency response and impedance measurements on two- and three-way systems using dynamic drivers plus a very interesting two-way system using a wide-band ribbon driver. Each example highlights some important aspect of the measurement process and gives practical advice on how to interpret and use the results. Several techniques for combining near-field low-frequency data with free-standing far-field or ground-plane measurements are presented. In some examples both the full system response and the response of the individual drivers are shown to illustrate the effect of the crossover network. Two of the examples show the set of measurements required to produce a crossover design and detail elements of the crossover design process.

In addition to on-axis measurements, the polar responses of several of the examples are presented. Polar response has a very strong impact on imaging and the perceived spectral balance in typical listening rooms. The polar response of a 60″ long wide-band ribbon driver is measured and shown to be cylindrical in nature. This polar response is contrasted with the spherical response pattern of dynamic drivers. The different response patterns are shown to complicate the integration of dynamic drivers with large linear arrays. The chapter ends with a brief discussion of power response illustrated with a subwoofer example.

The frequency source for all *acoustic* measurements in this chapter is a sine-wave oscillator frequency modulated to produce 1/3-octave-wide warble tones. A calibrated laboratory microphone, together with a wideband preamp, is also used in all of the tests described. The amount of system information realized from this simple instrumentation suite is quite surprising.

1.4.5 CHAPTER 6: TIME, FREQUENCY, AND THE FOURIER TRANSFORM

With the advent of PC-based electrical and acoustical data-acquisition and analysis systems, you are now able to quickly and accurately measure many more driver and loudspeaker system parameters. These include transient behavior in the time domain, inter-driver time delay, phase response, and driver acoustic phase centers, to name a few.

In the preceding chapters all of the measurement techniques described and all of the examples presented involved analog measurements in the frequency domain. PC data-acquisition systems work in the time domain. They are the digital counterparts of an oscilloscope, capturing and displaying waveforms on a time axis. Fortunately, for linear systems, there is a direct relationship between time-domain data and frequency-domain data. For a loudspeaker the equivalent of frequency response in the frequency domain is impulse response in the time domain.

Impulse response and frequency response are related by a mathematical operation called the Fourier transform. It is important to have a strong qualitative understanding of the Fourier transform and many additional concepts in digital signal processing and linear systems to properly interpret and use the data these systems provide.

To this end, the impulse signal is introduced and a loudspeaker's impulse response is defined with an example. The frequency response corresponding to this impulse response is also shown. There is a problem, however. Loudspeaker impulse response is a continuous-time function and frequency response is a continuous-frequency function. That is, the impulse response has a value for every point in time, and the frequency response has a value for every point in frequency. But computers must work with samples of either process, known only at discrete instants in time or at discrete points in frequency.

The next five sections of Chapter 6 explain how to get a good representation of impulse response and frequency response from sampled data. This involves the sampling of continuous-time signals and elements of digital signal processing fundamental to the operation of PC-based systems.

In Section 6.3 continuous-time periodic signals are defined and several examples of these signals are given. The Fourier series representation of continuous-time periodic signals is then presented. This series is shown to comprise an infinite sum of harmonically related sinusoidal signals. This repre-

sentation supports the conclusion that a periodic signal has a discrete-frequency spectrum.

Next I describe the sampling process, which converts a continuous-time signal into a discrete-time signal whose value is known only as discrete, equally spaced instants in time. Then I present the sampling theorem, which sets an upper bound on the frequencies which can be unambiguously recovered by the sampling process. The problem of frequency aliasing is then explained with examples.

The Discrete Fourier Transform (DFT) is presented in Section 6.4. This is basically a Fourier series for sampled data. The DFT is periodic. This leads to the problem of frequency leakage in the DFT when the time-sampled sequence itself is not periodic over the sample interval. A partial solution to spectral leakage using data windows is discussed and illustrated with examples.

The Fast Fourier Transform (FFT) is a computationally efficient algorithm for computing a restricted version of the DFT. The FFT is discussed in Section 6.5. Under the right conditions, the FFT of a time-sampled impulse response is shown to be a sampled version of the continuous-frequency frequency response. This is an extremely important result. The conditions for which this is true are explained.

At this point the connection between the sampled impulse response and the continuous-frequency frequency response which was sought has been established. The section continues with a discussion of FFT frequency resolution and presents a technique for producing smoother FFT plots with zero padding.

A practical technique for measuring loudspeaker impulse response using a Maximum Length Sequence (MLS) pseudo-random noise is introduced in Section 6.6. The MLS signal is described and advantages of the approach are listed. The low crest factor of the MLS signal guarantees linear operation of the device under test (DUT) while providing a greatly increased signal-to-noise ratio relative to other techniques. A block diagram overview of the MLS signal processing is also presented.

At this point you have all the concepts from Fourier theory and digital signal processing you need to understand the operation of PC-based data-acquisitions systems. The chapter now turns to an explanation of the many new driver and loudspeaker performance variables you can now measure.

In Section 6.7 phase and phase response are precisely defined. Examples are used to show that phase angles in excess of a few hundred degrees are possible in many networks and loudspeaker systems. The periodicity of sine waves in phase, however, limits phase response measurements to $\pm 180°$. The process of unwrapping measured phase to get true phase is explained with an example.

The concepts of a minimum phase response and excess phase are then introduced. This leads to a description of all-pass filters and the phase response of a pure time delay. Group delay is also defined and examples presented.

Section 6.8 begins with a simple example showing that there is no single point in space or time characterizing the acoustic spatial or temporal location of a driver. In regions where frequency response is changing, the signal source locations for even a single frequency are distributed over a small region in space or, equivalently, over time.

Fortunately, all is not lost. The discussion of phase in Section 6.7 is now used in Section 6.8 to define the acoustic phase center of a driver. This is the proper characterization of a driver's acoustic location needed in crossover network design to assure proper multiple-driver integration. Using the minimum phase property of drivers, a technique for finding a driver's acoustic phase center is illustrated with an example.

1.4.6 CHAPTER 7: LOUDSPEAKER TESTING WITH PC-BASED ACOUSTIC DATA-ACQUISITION SYSTEMS

Chapter 7 is devoted entirely to examples of measurements made with the MLSSA and CLIO PC-based electrical and acoustical data-acquisition and analysis systems. The chapter starts with a brief description of the capabilities of each system. The primary stimulus for MLSSA measurements is the MLS sequence. CLIO also has an MLS signal, but additionally it provides pink noise, sine waves, swept sine waves, tone bursts, and multitone signals.

The chapter continues with examples of impedance and capacitance and inductance measurements made using both systems. CLIO and MLSSA use the voltage-divider approach for impedance measurement, but because these systems measure phase in addition to magnitude, the limitations on the voltage-divider technique imposed in Chapter 2 are removed and exact measurement of impedance magnitude and phase is obtained.

CLIO uses sine waves for its impedance measurements, while MLSSA results are obtained with the MLS signal. In spite of the different test stimuli, examples show that both systems produce comparable results with typical circuit elements. Phase data examples are used to show the nonideal behavior of real world inductors and capacitors.

Measurement of T/S parameters with MLSSA and CLIO is discussed next. The T/S parameters are derived from impedance data. However, both CLIO and MLSSA improve upon the "three-point" method discussed in Chapter 2 by fitting a voice-coil impedance model to the measured voice-coil impedance. T/S parameters are then computed from the model parameters.

Errors produced by impedance curve asymmetry and measurement noise are greatly reduced with this approach. T/S parameters of the same driver measured with both systems are shown to yield small, but significant, differences due to the effective drive level produced by the different stimuli and the fact that T/S parameters are drive-level dependent. A second example shows that results can be made approximately equal with proper setting of the MLS drive level relative to the sine wave amplitude.

The chapter continues with a discussion of the more commonly used measurements provided by MLSSA and CLIO for loudspeaker design and evaluation. The measurement types fall into two general categories: time domain and frequency domain. Time-domain measurements include the impulse response, step response, the energy-time curve (ETC), and the cumulative spectral decay (CSD).

Step response is shown to be a useful qualitative measure of a loudspeaker system's time coherence. The ETC is very helpful in evaluating the arrival of reflections in reverberant spaces. The CSD measures the spectrum of decay energy in a loudspeaker following an impulsive input. The CSD reveals resonant modes that hang on after a signal is removed. Examples of each measurement type are presented.

Reflections from surfaces such as the floor, ceiling, and side walls appear at different times in the impulse response. By removing portions of the impulse response containing reflections, it is possible to obtain anechoic driver and loudspeaker frequency response over a limited frequency range. An example here shows the effect of including specific reflections in the frequency-response calculation.

Frequency-domain measurements include frequency response, phase response, and driver acoustic phase center. Additionally, MLSSA supplies excess phase, group delay, and excess group delay. Excess group delay is shown by example to provide a very accurate measure of inter-driver time offsets in loudspeaker systems. With its sine-wave sources and FFT spectrum analyzer, CLIO will measure harmonic and intermodulation distortion.

The measurement capabilities described above are now applied in two distinct areas: loudspeaker system design and loudspeaker system performance evaluation. The measurements needed in each area are somewhat different. With regard to the design process, modern crossover optimization software packages require very detailed information on driver impedance, frequency and phase response, and acoustic phase center location before the optimization process can begin. Design data must be specific to the acoustic environment in which the system of drivers will operate.

The following data is needed for the design process: voice-coil impedance magnitude and phase for each driver, driver on-axis frequency and phase response, and the relative location of driver acoustic phase centers. A two-way design example shows the full data-collection process and follows the design through to completion. All intermediate measurement results and the final system response are shown.

The measurements needed for loudspeaker system performance evaluation are different from, and more extensive than, those required for design purposes. Within their limitations, measurements for performance evaluation should help you to determine how a system will sound when placed in a typical listening environment.

The measurements that in my experience relate most directly to this goal are: system impedance magnitude and phase, on-axis frequency response, system sensitivity, cumulative spectral decay, step response, excess group delay, horizontal and vertical polar response, power response, and harmonic and intermodulation distortion. The text reports the results of this full sequence of these tests performed on a two-way MTM system. The results of each test are shown in graphical or tabular form, and sonic implications of the test results are carefully explained.

The chapter ends with three additional examples that illustrate the broad utility of PC-based electrical and acoustical data-acquisition systems. In the first example, excess phase response is used to compute the sound velocity in a stuffed transmission line. The second example takes another look at data windows and spectral leakage. The third and final example examines a very special window and its use in room/loudspeaker frequency-response analysis and equalization.

I hope this brief introduction to *Testing Loudspeakers* has sparked your interest in the subject. If so, please go on to Chapter 2.

REFERENCE

1. Dickason, Vance, *The Loudspeaker Design Cookbook*, Fifth Edition, Audio Amateur Press (PO Box 243, Peterborough, NH 03458).

CHAPTER 2

DRIVER TESTING

2.0 INTRODUCTION

The two most important data sets describing a loudspeaker driver are its frequency response (both magnitude and phase) and its *Thiele/Small (T/S) parameters*. Driver frequency response measurements will be discussed in Chapter 4. The T/S parameters of a driver are needed to correctly design an enclosure.

In this chapter I will give a very brief historical development of low-frequency direct-radiator loudspeaker system design, which highlights the introduction of the T/S approach, define the T/S parameters, give a brief development of their meaning from physical principles, and present techniques for measuring them. Because the driver impedance curve plays a major role in determining T/S parameters, its nature will also be discussed from first principles. Use of the T/S parameters in low-frequency direct-radiator loudspeaker system design is well covered in the *Loudspeaker Design Cookbook (LDC)*.[1] The reader is assumed to be familiar with this work.

2.1 A LITTLE HISTORY

From the invention of the electrodynamic loudspeaker by Kellogg and Rice[2] (circa 1920) until the 1970s, direct-radiator loudspeaker system design was as much an art as a science. The critical relationships between driver parameters, enclosure design, and loudspeaker system low-frequency response were poorly understood. A direct-radiator loudspeaker is a complex device first involving a transformation of electrical energy to mechanical energy followed second by a conversion of mechanical energy to acoustic energy. Although equations describing this multistep process were developed in the 1930s (e.g., McLachan[3]), these equations were complex and extremely difficult to solve. General solutions suitable for design purposes were not available.

In the 1940s, Olson[4] and others introduced the concept of the analogous electric circuit to the low-frequency analysis of acoustic systems. In this approach acoustical elements are modeled as combinations of resistors, inductors, and capacitors. This is possible because under certain reasonable assumptions the *low-frequency* behavior of the acoustical elements obeys the same equations as their corresponding electrical elements. This meant that a complex system of acoustic elements could be modeled as an electric circuit and all the power of a highly developed electric circuit theory could be brought to bear on the analysis of loudspeaker system performance. Complete electric circuit models for closed-box and vented loudspeaker systems were given by Olson[5] in 1943.

Unfortunately, as promising as the analogous circuit approach was, a big problem still remained. Some of the acoustical elements are frequency dependent. This means that the value of the analo-

gous electrical element must change with frequency. This greatly complicated the analytic solution of the circuit equations, and numerical solutions were out of the question without modern computers. Much of the frequency dependence could be ignored, but it was widely believed that one critical parameter, *radiation impedance*, could not. Radiation impedance, and in particular, the radiation resistance of a driver cone, is the parameter that relates cone velocity and acoustic output power. Ignore radiation resistance and you have no sound!

Using Olson's analogous circuit for the bass-reflex system, Beranek,[6] in 1954, was able to solve the bass-reflex system response equations at three critical points where the radiation resistance could be calculated, but a complete response curve was still lacking. A major breakthrough in loudspeaker system analysis came in 1959.

Recognizing that direct-radiator loudspeaker systems are very inefficient (typical efficiencies are 1% or less), Novak[7] reasoned that the effect of radiation resistance loading on the system could be ignored in the calculation of cone velocity. Once the velocity is known, acoustic output could then be calculated using the frequency-dependent radiation resistance. This led to a relatively simple mathematical solution for frequency response. Novak then analyzed the response of closed-box and vented systems and presented an analysis of the relative performance of the two.

An engineering discipline is not truly mature until it moves from analysis to synthesis. Given a set of system performance specifications, analytical tools allow a designer to converge on the correct design through a series of trials. After each trial, driver and/or enclosure parameters are changed and the analysis repeated until the desired performance is attained. Starting with a set of system performance requirements, synthesis tools lead directly to the required design. The birth of the synthesis approach to direct-radiator loudspeaker design is generally attributed to A.N. Thiele.

In 1961 he published a landmark paper entitled "Loudspeakers in Vented Boxes."[8] Building upon Novak's work, Thiele showed how to select driver parameters, enclosure volume, and tuning so as to match or *align* a vented-box system-frequency response with that of a number of well known electrical filter responses. Since frequency, phase, and transient response of the electrical filter are known, it followed that a loudspeaker aligned to a particular filter would have the same response. Thiele provided a table of nine discrete alignments for the vented enclosure.

He also presented some relatively simple procedures for measuring the primary driver parameters (now called Thiele/Small parameters) needed for the design process, namely resonant frequency, f_S, the Q, and equivalent volume, V_{AS}, using only driver electrical impedance data. With Thiele's paper, the loudspeaker system designer now had a systematic procedure for specifying driver parameters and enclosure design (volume and tuning) for vented systems to achieve a desired response.

At this point we might think that the analysis and synthesis of direct-radiator loudspeaker systems was well-in-hand, but publications in professional society technical journals often trickle down slowly into the design community. Thiele's paper was published in an Australian technical journal and it was several more years before his results were widely known.

Over the period 1968–1972, J.E. Benson published a series of papers building on the works of Novak, Thiele, and many others, entitled, "Theory and Design of Loudspeaker Enclosures." [9] This three-part series of some 240 pages provided a very general theory for all closed, vented, and aperiodically damped enclosures. Unfortunately, Benson's papers were also published in an Australian journal and languished for many years until they were referenced in Small's papers, discussed next.

The event that brought low-frequency, direct-radiation loudspeaker system synthesis to the attention of the general design community was the publication of R.H. Small's papers,[10-12] in the *Journal* of the Audio Engineering Society (*JAES*). Small, Benson's Ph.D. student, published a series of nine articles on the analysis and design of closed-box, vented, and passive-radiator systems. These papers expanded upon Thiele's work, providing detailed proofs of many of his results and a wide range of design examples. Small introduced the *continuous alignment graph* which replaced Thiele's limited discrete alignment table. With this plot a designer could find the proper enclosure volume (and tuning for vented systems) for a continuous range of driver parameters. Because the *JAES* had international distribution, Small's results were widely disseminated and by the mid-'70s, many designers were applying his design methodology with great success.

2.2 THE THIELE/SMALL PARAMETERS

Almost all of the useful driver parameters had been defined by other researchers and authors before Thiele and Small. However, they brought these parameters and a few new ones together in a complete design approach and showed how they could be easily determined from impedance data alone. The parameters are now known collectively as the Thiele/Small parameters.

Figure 2.1 presents a cutaway view of a generic low-frequency driver, i.e., a woofer. This figure is taken from Reference 1 and should be familiar to the reader. With reference to this diagram, a partial list of T/S parameters is given below. Their meaning and physical significance will be discussed a bit later in the chapter.

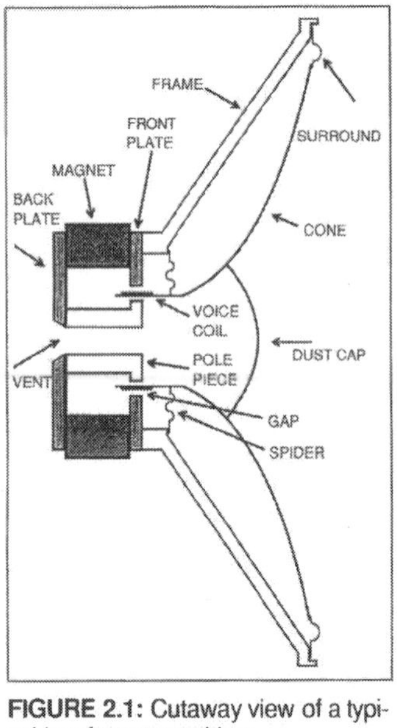

FIGURE 2.1: Cutaway view of a typical low-frequency driver.

The T/S parameters of primary interest, listed in alphabetical order, are:

B	Magnetic flux density in the driver gap
Bl	Motor strength—the product of B times l
C_{AS}	Acoustic compliance of the driver suspension
C_{MS}	Mechanical compliance of the driver suspension
f_S	Resonant frequency of driver including air load
f_{SA}	Free-air resonant frequency of driver
l	Length of voice coil wire in the magnetic field gap
L_{VC}	Voice coil inductance
M_{AS}	Acoustic mass of driver cone assembly including reactive air load
M_{MD}	Mechanical mass of the driver cone assembly, excluding air load
M_{MS}	Mechanical mass of diaphragm cone assembly including air load
Q_{ES}	Q of driver at f_S considering electrical resistance only
Q_{MS}	Q of driver at f_S considering mechanical losses only
Q_{TS}	Q of driver at f_S considering all driver losses
R_{AS}	Acoustic resistance of driver suspension losses
R_E	DC electrical resistance of voice coil
R_{ES}	Reflected electrical resistance due to driver suspension losses
R_{MS}	Mechanical resistance of driver suspension losses
S_D	Effective surface area of the driver cone
S_0	Reference voltage sensitivity of the driver
S_p	Reference *power sensitivity* of the driver
V_{AS}	Volume of air having the same acoustic compliance as driver suspension
η_0	Reference efficiency of driver

In the above list, the letter A used in the first position of the subscript refers to acoustic parameters. Electrical and mechanical parameters are denoted by E and M, respectively.

2.3 THE LOUDSPEAKER DRIVER IMPEDANCE CURVE

As a complex electro-mechano-acoustic transducer, a loudspeaker driver is characterized by a mix of electrical, mechanical, and acoustic parameters, many of which were listed in the previous section. The beauty of the T/S theory is that all driver parameters needed for loudspeaker system design can be determined from the *driver impedance curve* measured under just two different conditions. That is, all parameters of interest can be determined from the electrical or input side of the driver. Really rather amazing when you stop to think about it.

Because of its importance, let's spend a little time understanding the driver impedance curve and where it comes from. *Figure 2.2* shows a typical low-frequency driver impedance plot. Both impedance magnitude and phase are shown. At very low frequencies the impedance magnitude is equal to the voice coil resistance, R_E, and its phase is close to 0°. The impedance rises with increasing frequency and the phase angle goes positive. In this region, below driver mechanical resonance, the driver electromotive, or motional, impedance looks inductive. The impedance curve then peaks at a value of $R_E + R_{ES}$ at the driver mechanical resonant frequency, f_S, where the phase goes through zero again.

Above resonance, the plot falls and the phase angle goes negative. In this region, the driver impedance looks capacitive. Next, the curve reaches a minimum, where the voice coil inductance, L_{VC}, forms a series resonance with the capacitive electromotive impedance of the driver. Finally, the impedance curve rises again due to voice-coil inductance and other frequency-dependent effects discussed a bit later in this chapter. These effects dominate driver impedance from this point on.

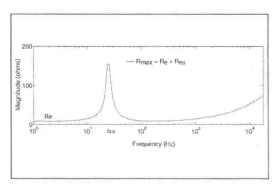

FIGURE 2.2a: Impedance magnitude of a typical 8″ driver.

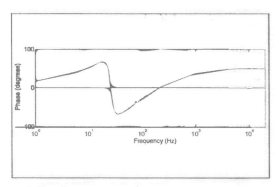

FIGURE 2.2b: Impedance phase of a typical 8″ driver.

2.4 A SIMPLE DRIVER MODEL
Why does the impedance curve look like the plot shown in *Fig. 2.2*? Let's build up a simple model for this woofer that we can use to explain the shape of the impedance curve.

2.4.1 MODELING THE MECHANICAL SIDE
Figure 2.3 depicts a weight of mass, M, suspended from a fixed surface; for example, a ceiling, on a spring of stiffness, K. The mass is specified in kilograms. K has the units of force over distance. For instance, if it takes a force of one newton (1N) to stretch the spring from its resting length by one meter (1m), then K has the value of one newton/meter (1N/m).

Figure 2.3 also shows a dashpot damping element, D, which has units of force over velocity or newtons per meter per second. That is, the damping force restraining motion of the mass, M, is proportional to the velocity of the mass. (In this book we will often use the Systeme Internationale

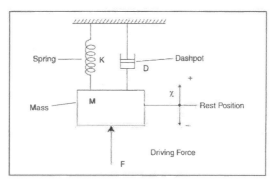

FIGURE 2.3: Mass/spring/dashpot model of driver mechanical properties.

(SI) set of units. The basic units of length, mass, and time in this system are the meter, kilogram, and second. The unit of weight or force is the Newton. Appendix A gives the conversion factors between Imperial and SI units.)

If the mass is pulled down by a fixed amount, potential energy is stored in the stretched spring. If the mass is now released it will bob up and down. In this idealized model the mass will bob up and down forever if D is zero; that is, if there are no mechanical losses in the system. The distance M moves from its rest or initial position–denoted by x(t) in the figure–will vary sinusoidally in time. This is easily seen by referring to *Fig. 2.4*.

If the attachment point of the spring at the ceiling is moved at a constant velocity and a light is shown on the moving mass, the mass's shadow will trace out a sine wave on the wall behind it, much like the trace on an oscilloscope. The amplitude of this harmonic motion will change in proportion to the size of the initial displacement, but the frequency or period, T, will not. The frequency will be the same regardless of how far the mass is initially displaced as long as the spring is not stretched beyond its elastic limit.

The mass/spring combination forms a harmonic oscillator. The frequency of oscillation is given by the equation:

$$f_S = \frac{1}{2\pi} \sqrt{\frac{K}{M}} \qquad [2.1]$$

where:

f_S = resonant frequency in Hz
K = spring stiffness in newtons/meter
M = mass in kilograms

Equation 2.1 tells us that a stiffer spring produces a

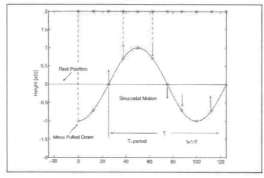

FIGURE 2.4: Periodic motion of spring/mass system.

11

higher resonant frequency, while a larger mass results in a lower f_S. This jibes well with intuition.

The concept of spring stiffness is intuitively appealing. In the loudspeaker world, however, it is customary to talk about spring or suspension *compliance*. Mathematically, compliance, C, is the inverse of stiffness.

$$C = \frac{1}{K} \qquad [2.2]$$

Thus a very stiff spring has a low compliance and vice versa. Compliance has the units of distance over force or meters/newton. In terms of compliance, Equation 2.1 becomes:

$$f_S = \frac{1}{2\pi} \sqrt{\frac{1}{CM}} \qquad [2.3]$$

M_{MD} is the total moving mass of the cone assembly. It includes the mass of the cone, voice-coil former and wire, the dust cap, and also a portion of the surround and spider which move with the cone assembly. There is an additional effective mass due to loading on the cone as it pushes against the surrounding air. This "air-load," referred to as mass reactance loading, will be discussed in more detail in Chapter 3. For now we simply note that the additional effective mass due to air load on a driver suspended in free air with no additional baffling is calculated with the following formula:

$$M_{M1} = 2.67\rho a^3 = 0.394D^3 = 0.566S_D^{1.5}$$

where:
  a = driver cone radius
  D = driver cone diameter
  ρ = density of air (1.18kg/m^3 at 20°C sea-level)

Then the total moving mass including air load becomes:

$$M_{MS} = M_{MD} + M_{M1}$$

tion for the free-air resonance becomes:

$$f_{SA} = \frac{1}{2\pi} \sqrt{\frac{1}{C_{MS}M_{MS}}} \qquad [2.4]$$

The damping, D, provides the mechanism for dissipating energy in the harmonic oscillator. If D is not zero, the oscillation excited by the initial displacement will decay to zero exponentially in time (*Fig. 2.5*). In addition to the decaying oscillation, *Fig. 2.5* shows two exponentially decaying curves, one above and one below the actual oscillation. These curves define an envelope that bounds the peak magnitude of the oscillation over time. The rate of decay of the envelope is described by the nondimensional parameter, Q_{MS}.

$$Q_{MS} = 2\pi f_S \left(\frac{M_{MS}}{C_{MS}} \right) = \left(\frac{1}{D} \right) \sqrt{\frac{M_{MS}}{C_{MS}}} \qquad [2.5]$$

Q_{MS} is inversely proportional to D. Thus, a large value of damping is equivalent to a low value of Q_{MS} and vice versa. The rate of decay of the oscillation in dB per cycle is:

$$DR = -54.6/Q_{MS} \quad dB/cycle: Q_{MS} > 0.5 \quad [2.6]$$

Equation 2.6 is valid for values of Q_{MS} greater than 0.5. Below this value the system is overdamped and an initial displacement decays uniformly to zero without oscillation. From Equation 2.6 it is clear that a system with a high Q will decay slowly in time, while a system with low Q will die rapidly. Q_{MS} ranges from 3–10 for commonly available drivers. Much more will be said about Q shortly.

2.4.2 THE FORCED SYSTEM

So far we have discussed what happens over time when the system of *Fig. 2.3* is disturbed by an initial displacement. What happens when the system is driven continuously with a sinusoidal force of constant amplitude and varying frequency? The displacement frequency response of the driven mass is shown in *Fig. 2.6a* for values of Q_{MS} of 0.5, 0.7, 1, 2.5, and 10. At frequencies much below resonance, the displacement response is constant. In this frequency region, the forcing motion is resisted only by the spring stiffness. This is referred to as the stiffness- or *compliance-controlled region*.

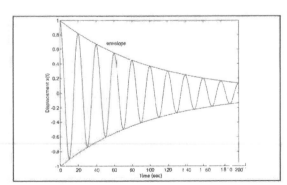

FIGURE 2.5: Decaying oscillation and its envelope.

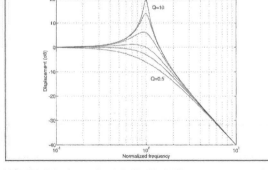

FIGURE 2.6a: Normalized displacement versus mechanical Q.

C_{MS} is the total compliance of the cone assembly and includes the compliance of the spider and surround. If we replace C and M in *Fig. 2.2* with C_{MS} and M_{MS}, we have a simple model for the mechanical portion of a driver, and the equa-

As the exciting frequency is increased, motion of the mass increases until it reaches a maximum at or slightly above f_S, depending on the value of Q_{MS}. From the graph it is clear that the peak of the response is controlled by the value of Q_{MS}, smaller values implying greater damping and less peaking. The region around f_S is said to be damping controlled. Somewhat above f_S response falls off, reaching a constant rate of $-12dB/octave$. Here inertial forces dominate as it becomes increasingly harder to move that mass with increasing frequency. This is called the *mass-controlled region.*

Figures 2.6b and *2.6c* give more insight into the frequency response of the system. *Figure 2.6b* depicts the velocity response of the driven mass. This curve has a bandpass characteristic, rising 6dB per octave at low frequencies, reaching a maximum at f_S and falling at 6dB per octave above f_S. This explains why the damping force, which is proportional to velocity, dominates in the frequency region around f_S, where velocity is highest. Unlike position response, velocity response is *always* maximum at f_S where velocity is highest. This fact will be used later to determine f_{SA} from the impedance curve.

Acceleration response is shown in *Fig. 2.6c*. Acceleration rises at 12dB/octave below f_S. Above f_S, however, acceleration flattens out and becomes constant. In order to keep the mass displacement response constant, acceleration must increase as the square of frequency.

Conversely, if acceleration is constant, displacement must fall as one over the square of frequency or 12dB/octave as was shown in *Fig. 2.6a*. Notice that the acceleration response plot is similar to the low-frequency response of a closed-box speaker system (see Fig. 1.1 of Reference 1). This is no coincidence because the pressure response of a loudspeaker is proportional to driver cone acceleration.

2.4.3 WHAT IS THIS THING CALLED Q?

In the early days of radio, parallel resonant LC tank circuits were used extensively to improve the frequency selectivity of radio-frequency receivers. (It is unfortunate that the symbol C is used for both mechanical compliance and electrical capacitance. The meaning of C should be clear by the context.) These tank circuits have a bandpass frequency response just like the velocity response of our spring/mass system. The narrower the peak in response, the more selective the circuit is.

The selectivity of the LC tank circuit is characterized by a nondimensional factor, the *quality factor*, or Q, of the circuit. Qs of 50, 100, or more were common for these circuits. A higher Q meant less damping or electrical loss in the circuit, and therefore, higher selectivity. Qs were limited by such factors as wire resistance in the coils and the dissipation factor of the capacitors of the day.

Q is often referred to as a measure of the energy lost by a system, but this is only half the story. Formally, Q is defined as a ratio:

Q = energy stored per cycle/energy dissipated
　　per cycle,

where "per cycle" refers to one cycle of the sine wave driving force, F. The driving force supplies the input energy to the system. Q is thus a measure of how much of the input energy is retained by the system versus how much is dissipated ultimately as

heat. If more energy is retained than lost, Q will be greater than one.

Clearly, this definition of Q can be applied to mechanical or electrical systems. For the mechanical system of *Fig. 2.2*, energy is stored alternately as potential energy in the compression or elongation of the spring, or as kinetic energy in the velocity of the mass. Energy is dissipated through the action of damping. In a driver, mechanical losses are produced by the flexing of the spider and surround. Later in this chapter, an electrical Q will be defined, which accounts for losses on the electrical side of the driver.

We now have a complete model for the mechanical side of a driver, so let's turn our attention to the electrical side.

2.4.4 MODELING THE ELECTRICAL SIDE

The voice coil together with the magnetic field in the voice coil gap form a linear motor. *Figure 2.7* is an expanded view of the voice coil and annular voice coil gap geometry. The magnetic flux lines are radial in the gap and perpendicular to the voice-coil windings. It follows that they are also perpendicular to any current flowing in the wire. When a current, i, is passed through the voice coil, a force, F, is produced and that drives the cone. The force is proportional to the magnetic flux density, B, the length of wire in the magnetic field, l, and the current, i. The appropriate equation is:

$$F = Bli \qquad [2.7]$$

The direction of the force is determined using the procedure illustrated in *Fig. 2.8a*. Draw an arrow in the direction of the magnetic flux lines. Draw a second arrow in the direction of the current flow. Rotate i into B through the smaller angle

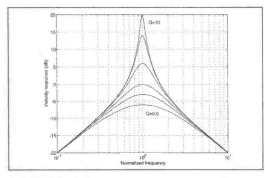

FIGURE 2.6b: Velocity response versus mechanical Q.

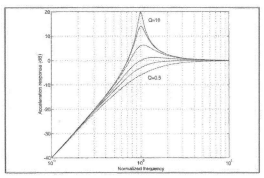

FIGURE 2.6c: Acceleration response versus mechanical Q.

13

between them. The force will be in the direction that a right-hand screw would move with that same rotation. For example, if i must rotate clockwise to meet B, a right-handed screw would advance into the plane of the paper. Known as the *right-hand rule*, it shows that the resulting force is always perpendicular to both the current direction and the flux lines; that is, it is always perpendicular to the plane formed by the i and B arrows.

For those of you familiar with vector algebra, the force F is computed as the vector cross product of i and B, where i is taken in the direction of the conductor, l. For a driver, the force acts along the voice-coil axis and reverses with reversal in the direction of the current. The linear motor pushes the voice coil and thereby the cone in and out relative to the driver frame.

At first glance it might appear that the voice-coil current is limited only by voice-coil impedance and the voltage applied to the coil terminals, but something else is going on here. The voice coil moving

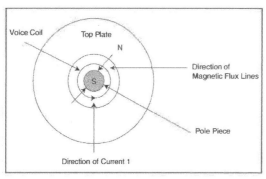

FIGURE 2.7: Voice-coil magnetic gap geometry. Force is into the paper for the direction of current flow and flux lines.

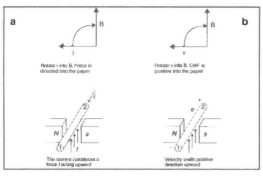

FIGURE 2.8: Illustrates the direction of force and back EMF.

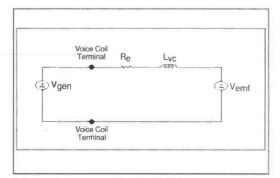

FIGURE 2.9: Equivalent circuit seen at voice-coil terminals.

through the magnetic field acts like a generator. According to one form of Faraday's Law, when a conductor of length, l, moves at right angles to a magnetic field, B, at a velocity, v, an electromotive force (EMF) or voltage is generated across the conductor of magnitude:

$$\text{EMF} = \text{B}lv \qquad [2.8]$$

The polarity of this voltage is shown in *Fig. 2.8b*. It is determined by drawing the arrows or vectors v and B and then rotating v into B through the smaller angle between them. The polarity of the back EMF again corresponds to the direction in which a right-hand screw would move. Notice that the polarity of the generated voltage causes a current to flow which opposes the current flow induced by the applied voltage, V_{gen}. For this reason it is referred to as a *counter*, or *back EMF*.

2.4.5 THE DRIVER IMPEDANCE CURVE

We are now in a position to explain the driver impedance curve. An equivalent circuit looking into the voice-coil terminals is shown in *Fig. 2.9*. Assume an AC sine-wave voltage of *constant amplitude*, V_{gen}, is applied to the voice-coil terminals (ignore voice-coil inductance for now). In response to the applied voltage, a current will flow in the voice coil, in turn generating a sinusoidally varying force, causing the cone to move back and forth. From Ohm's Law, the current, i_{VC}, at very low frequencies will equal:

$$i_{VC} = \frac{V_{gen}}{R_E}$$

As the sine-wave frequency is increased, however, cone velocity will increase according to *Fig. 2.6b*, reaching its maximum value at f_{SA}. Likewise, the back EMF producing the opposing current flow will increase and thus the net current in the voice coil will decrease to a minimum at f_{SA}. As frequency is further increased, velocity and back EMF will fall off, ultimately allowing the current to return to its low-frequency value (remember L_{VC} is ignored for the moment). A typical current versus frequency profile is illustrated in *Fig. 2.10a*.

For a different look at the driver impedance function, now assume that the applied voltage amplitude is varied with frequency so as to main-

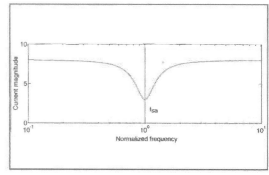

FIGURE 2.10a: Voice-coil current with constant-voltage sweep.

tain a *constant current* in the voice coil. Again from Ohm's Law, the current, i_{VC}, is:

$$i_{VC} = \frac{V_{gen} - V_{emf}}{R_E} \qquad [2.9]$$

As frequency is increased, the back EMF will again increase, and therefore, applied voltage (V_{gen}) must increase to keep the current constant. The applied voltage will reach a maximum at f_{SA} and fall off as frequency is increased beyond f_{SA}. A plot of the applied voltage amplitude versus frequency is shown in *Fig. 2.10b*.

Since voice-coil impedance is defined as V_{gen}/i_{VC}, it should be clear from the previous discussion and *Fig. 2.10* that voice-coil impedance reaches a maximum at f_{SA}, assuming L_{VC} is zero. Back EMF is the mechanism by which the mechanical and acoustical properties of a driver are reflected back into the electrical impedance seen at the voice-coil terminals. Thiele and Small showed us how to extract the T/S parameters from the impedance data.

The impedance rise in the vicinity of f_{SA} is identical to that of a parallel resonant RLC circuit. It follows that an analogous all-electrical circuit model for voice-coil impedance can be constructed. *Figure 2.11* gives the complete electrical model for driver impedance considering all electrical and reflected mechanical and acoustic effects. In *Fig. 2.11* the *electrical capacitance*, C_{MES}, represents the total moving mass of the driver—including air load—reflected back to the electrical side. (See Chapter 3 for a discussion of the difference between M_{MS} and M_{MD}.) C_{MES} is related to the moving mass by:

$$C_{MES} = \frac{M_{MS}}{(Bl)^2} \qquad [2.10]$$

L_{CES} represents the driver's *mechanical compliance* reflected back to the voice-coil terminals and is given by:

$$L_{CES} = C_{MS}(Bl)^2 \qquad [2.11]$$

Finally, R_{ES} represents the *mechanical losses* reflected back to the electrical side. It is computed by:

$$R_{ES} = \frac{(Bl)^2}{R_{MS}} \qquad [2.12]$$

Notice in Equations 2.10–2.12 that all conversions of mechanical parameters to the electrical side involve the factor $(Bl)^2$. The first "Bl" comes from the "Bli" effect while the second one comes from the "Blv" term. Notice also that the complete circuit includes frequency-dependent resistance and inductance indicated in the drawing as R_{fd} and L_{fd}. That is, both the resistive and inductive portions of the equivalent circuit vary with frequency.

Audio currents passing through the voice coil produce an AC magnetic field, which in turn induces circulating currents in the pole piece, called *eddy currents*. The eddy currents generate heat that increases with increasing frequency. The net result is an increasing resistance and a decreasing inductance as the applied frequency rises. This effect will be discussed in more detail later in this chapter.

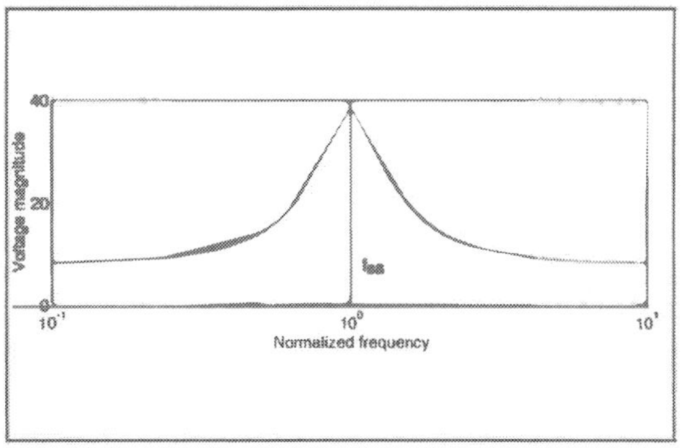

FIGURE 2.10b: Voice-coil voltage with constant current sweep.

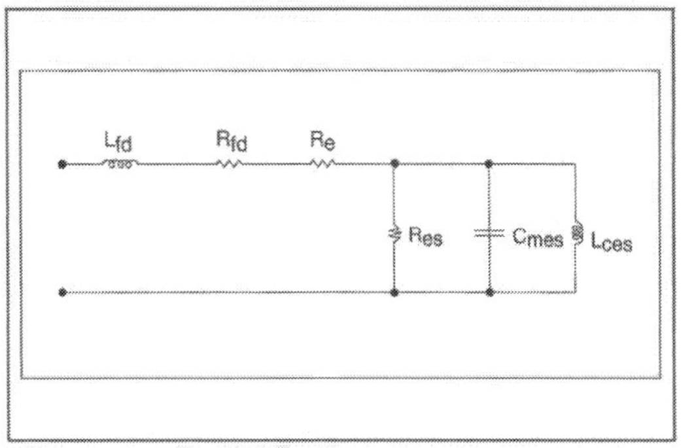

FIGURE 2.11: Equivalent electric circuit for driver impedance.

2.5 A BIT MORE ON T/S PARAMETERS

Before getting to actual parameter measurement techniques, two more parameters need to be defined and their significance explained. They are Q_{ES} and V_{AS}.

2.5.1 BACK EMF, ELECTRICAL DAMPING, AND Q_{ES}

Back EMF produces electrical damping which augments mechanical damping in a driver. In fact, for drivers commonly used in direct-radiator loudspeakers today, electrical damping is often the dominant source of damping, exceeding that from mechanical and acoustical losses.

To see how back EMF works, try the following experiment. Take a large, high-compliance woofer, preferably a 12″ or 15″ unit with a low f_S. Press down sharply on the cone near the dust cap. The cone will move easily. Now short the voice-coil terminals and again press down sharply. You will feel a much greater resistance to motion. Pressing down sharply on the cone imparts a high velocity to the voice coil, which in turn generates a back EMF.

This is easily observed by placing a voltmeter across the open voice-coil terminals. With the terminals shorted, however, the back EMF produces a large current in the voice coil, and from Equation 2.7, a force which opposes the cone motion and thereby damps the cone. Loudspeakers are usually driven from low output impedance amplifiers in order to maximize the effect of electrical damping.

Thiele defined an *electrical Q*, Q_{ES}, to characterize this electrical damping. The equation for Q_{ES} is:

$$Q_{ES} = \left(\frac{R_E}{(Bl)^2}\right)\sqrt{\frac{M_{MS}}{C_{MS}}} \qquad [2.13]$$

Physically, this equation tells us that Q_{ES} is inversely proportional to the square of the magnetic field strength so that a bigger magnet means more electrical damping. Q_{ES} increases with increasing R_E because a larger value of R_E reduces the current flow in the voice coil when its terminals are shorted and this, in turn, reduces the damping force. Q_{ES} increases with increasing cone mass simply because it is harder to damp a more massive cone. Finally, Q_{ES} decreases with increasing compliance because it takes less force to damp a more compliant suspension.

2.5.2 AIR SPRINGS, MECHANICAL COMPLIANCE, AND V_{AS}

Figure 2.12 shows two sealed boxes. Each box is equipped with a piston of area S_D forming an air spring. The second box has the smaller volume. If both pistons are pushed down the same distance, a greater force is needed to push down piston 2. Since we have defined a spring stiffness, K, as the

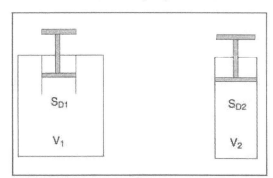

FIGURE 2.12: Effect of enclosed volume on air-spring compliance. V_1 is greater than V_2, but S_D equals S_{D2}. For this condition C_{M1} is greater than C_{M2}.

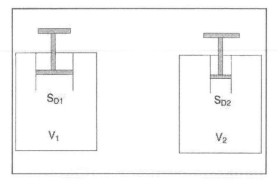

FIGURE 2.13: The effect of piston size on air-spring compliance.

ratio of force/distance, it is clear that air spring 2 is stiffer than air spring 1. Conversely, spring 1 is more compliant than spring 2. For a fixed piston area, the larger the enclosed volume, the higher its compliance becomes.

Figure 2.13 shows two boxes of equal volume, but now the piston in box 2 is smaller than that in box 1. If both pistons are again pushed down the same distance, a greater volume of air will be compressed by piston 1 relative to piston 2, and the force required to move piston 1 will be greater than that for piston 2. In this example, spring 2 is more compliant than spring 1. *We see that air-spring compliance depends on both the enclosed volume and the piston area.*

The air spring analogy can be used to specify a driver's compliance in terms of an equivalent volume of air. The appropriate parameter (V_{AS}) represents the enclosed volume of air which has the same compliance (or stiffness) as the suspension system of the driver when that air volume is compressed by a piston of the same area as the driver diaphragm. V_{AS} is a useful concept because it helps us to visualize the range of enclosure volumes in which a particular driver will work well. Specifically, V_{AS} is used to calculate the enclosure volume needed to obtain a particular response alignment using the α parameter described in Chapter 1 of *LDC*. The relationship between mechanical compliance (C_{MS}) and V_{AS} is given by the formula:

$$V_{AS} = \rho c^2 C_{MS} S_D^2 \qquad [2.14]$$

where:
ρ = density of air (1.18 kg/m^3 at 20°C sea-level)
c = speed of sound (344.5ms at 20°C sea-level)

V_{AS} is an acoustic parameter. For a given mechanical compliance, the *acoustic volume*, V_{AS}, varies as the square of S_D. This explains why large-diameter high-compliance drivers need large enclosures. Also note that V_{AS} changes with changes in air temperature and atmospheric pressure because both ρ and c change with these variables.

2.6 DETERMINING BASIC T/S PARAMETERS FROM THE IMPEDANCE CURVE

As an absolute minimum, the parameters f_{SA}, Q_{TS}, and V_{AS} are required to design an enclosure for a driver. Measurement of these parameters will be discussed first. Thiele recommended that the tested driver be suspended in air as far from any sound-reflecting surfaces as possible.

The resulting measured parameters are called the *"free-air" parameters*. Thiele (and later Small) derived a series of equations for determining the T/S parameters from the simplifed model of voice-coil impedance. We have already stated that radiation resistance can be ignored. Thiele also ignored the effect of voice-coil inductance, which is usually small in the frequency region around resonance. With these assumptions, a typical voice-coil impedance magnitude plot is depicted in *Fig. 2.14*.

Several features stand out in this figure. The

impedance magnitude plot reaches maximum value, R_{max}, equal to $R_{ES} + R_E$ at f_{SA} and the phase is zero; that is, the impedance is purely resistive at f_{SA}. Notice that the impedance magnitude returns to R_E at both very low and very high frequencies. Finally, the impedance magnitude plot is symmetric about f_{SA} when plotted on a logarithmic frequency scale. This last observation implies that the curve has geometric symmetry about f_{SA} and for any two frequencies, f_1 and f_2, symmetrically placed about f_{SA} on a logarithmic scale we have:

$$f_1 f_2 = f_{SA}^2 \qquad [2.15]$$

This fact can be used later in testing to verify the accuracy of Thiele's simplifying assumptions.

Starting from *Fig. 2.14*, Thiele developed the following relationships:

First define:

$$r_0 = \frac{R_{max}}{R_E} = \frac{R_{ES} + R_E}{R_E} \qquad [2.16]$$

and

f_1, f_2 = frequencies below and above f_{SA}, respectively, where the impedance magnitude, R_x, equals $R_E \sqrt{r_0}$.

Then:

$$Q_{MS} = \frac{f_{SA}\sqrt{r_0}}{f_2 - f_1} \qquad [2.17]$$

and

$$Q_{ES} = \frac{Q_{MS}}{r_0 - 1} \qquad [2.18]$$

$$Q_{TS} = \frac{Q_{MS} Q_{ES}}{Q_{MS} Q_{ES}} = \frac{Q_{MS}}{r_0} \qquad [2.19]$$

From these equations practical techniques for measuring f_{SA} and Q_{TS} can be developed. Referring to *Fig. 2.14*, we must measure R_E and three points on the impedance curve to determine f_{SA} and Q_{TS}. Three approaches will be discussed: the *voltage-divider procedure*, the *constant-voltage technique*, and the *constant-current technique*.

The voltage-divider approach uses the simplest instrumentation suite, but can be less accurate if not properly corrected. The constant-voltage procedure is more accurate, but requires a power amplifier. The constant-current technique uses an AC constant-current source (CCS). It is as accurate as the constant-voltage approach and somewhat quicker and easier to use, but requires the most sophisticated instrumentation.

2.6.1 TESTING PRELIMINARIES
The suspension on a driver will loosen up with

use, and therefore, its parameters will shift. In order to account for this, it is important that all drivers be broken-in before testing. This can be accomplished by suspending the driver in free air and driving it with a power amplifier at a frequency in the 20–25Hz range. The drive level should be set for moderate cone travel. Be careful not to overdrive the speaker or physical damage may result. Break in the driver for at least one hour. Longer periods may be needed. Some manufacturers specify their driver parameters after eight and even 24 hours break-in.

The T/S parameters are "*small signal*" parameters. It is important in all the tests described below to keep drive levels as low as your instrumentation will allow while still providing reliable results. When suspended in air, a driver cone has little resistance to motion at low frequencies. Under this condition, relatively small drive levels can produce large cone/voice-coil excursions which push the driver suspension or motor into a nonlinear region. This is especially true when the driving frequency is at or near driver free-air resonance. T/S parameters measured under large excursions will differ from their "small signal" value.

2.6.2 THE VOLTAGE-DIVIDER TECHNIQUE
The voltage-divider technique is diagrammed in *Fig. 2.15*. The following equipment is required:

1. Audio frequency sine-wave signal generator
2. High-impedance AC voltmeter
3. Frequency counter with 0.1Hz resolution
4. 1kΩ resistor (5W rating)
5. 10Ω, 1% resistor (see discussion below)
6. X,Y oscilloscope (optional)
7. Driver being tested

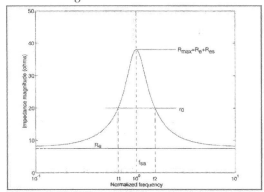

FIGURE 2.14: Voice-coil impedance magnitude (simplified model).

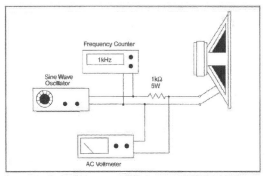

FIGURE 2.15: The voltage-divider test setup.

2.6.2.1 THE VOLTAGE DIVIDER PRINCIPLE

Figure 2.16 shows a simple voltage divider. The voltage at the output terminals of the divider is given by the following equation:

$$V_{out} = \left(\frac{Z}{R + Z}\right) V_{gen} \qquad [2.20]$$

The input voltage is "divided" down by the ratio $Z/(R+Z)$. Z is an arbitrary impedance which in general will have both resistive and reactive components. If R is much larger than the magnitude of Z—denoted by vertical bars—Equation 2.20 will simplify to:

$$V_{out} = \left(\frac{|Z|}{R}\right) V_{gen} = \left(\frac{V_{gen}}{R}\right)|Z|$$

or solving for $|Z|$,

$$|Z| = \left(\frac{R}{V_{gen}}\right) V_{out} \qquad [2.21]$$

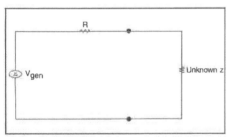

FIGURE 2.16: Voltage divider.

If V_{gen} is kept constant as frequency is varied, R/V_{gen} will be constant and $|Z|$ will be directly proportional to V_{out} as long as R is much larger than $|Z|$. The proportionality constant is R/V_{gen}.

To calibrate the voltage divider for impedance measurements, replace the loudspeaker being tested with the 10Ω, 1% resistor and adjust the output voltage of the signal generator until the voltage across the test resistor reads 0.01V (10mV). This sets the ratio between the calibration resistor and the measured voltage at 1000:1 and provides a convenient readout scaling of voltage to impedance.

If the calibration resistor is now replaced with an arbitrary impedance, the value of that impedance is determined simply by multiplying the voltage reading by 1,000. For example, a reading of 0.018V represents an impedance of 18Ω, 0.0056V equals 5.6Ω, and so forth. The ratio 0.01V = 10Ω was chosen because it is so easy to multiply by 1,000. Other ratios may be used if desired.

If a 10Ω, 1% resistor is not available, you may use any low-value precision resistor in its place. For example, if you have an 11.8Ω, 1% resistor, simply adjust the meter to read 0.0118V and the 1000:1 ratio is attained. Alternatively, you can measure any low-value resistor with a DC bridge or a digital multimeter (DMM) to 0.1Ω and use it as the calibrating resistor. Just remember to calibrate out the DMM lead resistance.

The voltmeter used to measure V_{out} must have an input impedance at least 100 times larger than the impedance under test. Wideband AC voltmeters such as the Hewlett-Packard HP403B rarely create a problem as they commonly have low-frequency input impedances of 1MΩ. Some older-style DMMs and multimeters with AC scales, however, may have rather low AC input impedances which can affect the accuracy of your results.

2.6.2.2 MEASURING F_{SA} AND Q_{TS} USING THE VOLTAGE DIVIDER

The voltage divider procedure is as follows:

1. Measure the DC resistance of the voice coil with a DMM or an accurate bridge. R_E should be measured at least to the nearest 0.1Ω. If a DMM is used, the lead resistance should be subtracted from your reading. The lead resistance can be determined by shorting the two leads together and recording the DMM reading in ohms.

2. Calibrate the voltage divider as discussed above.

3. *Keeping the signal generator output voltage constant*, vary the frequency to find the maximum impedance. Record the maximum impedance and label it R_{max}. Also record the frequency at R_{max} and label it f_{SA}.

4. Calculate r_0 using Equation 2.16 and then calculate R_x, where:

$$R_x = R_E \sqrt{r_0} \qquad [2.22]$$

5. Find the two frequencies, f_1 and f_2, respectively, below and above f_{SA} where the impedance is equal to R_x.

6. Check the validity of the simplified impedance model by calculating:

$$f'_{SA} = (f_1 f_2)^{1/2},$$

and compare the result to f_{SA}. If f'_{SA} is within 1Hz or less of f_{SA} for woofers, or 1% or less for midrange and tweeter drivers, the impedance model assumptions made in deriving the T/S parameter equations are valid. Larger differences generally imply large voice-coil inductance or non-linear effects due to excessive drive voltage.

7. If the test of step 5 is passed, driver Qs can be calculated as follows:

$$Q_{MS} = \frac{f_{SA} \sqrt{r_0}}{f_2 - f_1}$$

and

$$Q_{ES} = \frac{Q_{MS}}{r_0 - 1}$$

and

$$Q_{TS} = \frac{Q_{MS} Q_{ES}}{Q_{MS} Q_{ES}} = \frac{Q_{MS}}{r_0}$$

2.6.2.3 A VOLTAGE-DIVIDER EXAMPLE
Example 2.6.1
The parameters of a stamped-frame 8″ woofer with a polypropylene cone and a butyl rubber surround were measured using the voltage-divider technique. Voice-coil DC resistance measured with a DC-resistance bridge was found to be:

$$R_E = 7.85\Omega$$

A 1W, 5% 1kΩ resistor was used for R. Its actual resistance measured 1,024Ω. The divider was calibrated with a 10Ω, 1% resistor and the test begun. The maximum impedance and the frequency at which it occurred were found to be:

$$R_{max} = 72.5\Omega \text{ and } f_{SA} = 32.3Hz$$

Then:

$$r_0 = \frac{72.5}{7.85} = 9.24$$

$$\sqrt{r_0} = 3.04$$

and

$$R_x = 3.04 \times 7.85 = 23.9\Omega$$

The frequencies corresponding to R_x were found to be:

$$f_1 = 22.1Hz \text{ and } f_2 = 48.6Hz$$

Checking the simplified impedance model assumptions we find:

$$f'_{SA} = (22.1 \times 48.6)^{\frac{1}{2}} = 32.8Hz,$$

which is well within 1Hz of the measured f_{SA}.

From these values, the Qs were computed as follows:

$$Q_{MS} = 3.04 \times 32.3/(48.6 - 22.1) = 3.71$$

$$Q_{ES} = Q_{MS}/(9.24 - 1) = 0.45$$

and

$$Q_{TS} = (3.71 \times 0.45)/(3.71 + 0.45) = 0.40$$

2.6.2.4 CAVEATS, PITFALLS, AND REMEDIES

(A) Many DMMs do not provide reliable resistance measurements at low resistance values. Ideally, R_E should be known to three figures. A DC resistance bridge may be a better choice for measuring R_E if your DMM is not reliable. A simple but accurate DC resistance bridge is described in Reference 15.

(B) The appeal of the voltage-divider technique is its simplicity. Once calibrated, impedance is easily read from the scaled voltage measurements. The technique, however, depends strongly upon the assumption that R is much larger than $|Z|$, the impedance to be measured. How good is the assumption?

Figure 2.17 shows the error incurred with the voltage divider when using a 1kΩ resistor for R and a 10Ω calibrating resistor. For impedances larger than 10Ω, the voltage divider *always* reads an impedance value *below* the true value. The error is not only a function of the impedance magnitude, but also the impedance phase angle.

Curves are shown for impedances with phase

angles of 0, ±45, and ±90°. These correspond to a pure resistance, an impedance with equal resistive and reactive components (either inductive or capacitive) and a pure reactance (again either inductive or capacitive). Notice that a pure resistance encounters the greatest error. When calibrated with a 10Ω resistor and testing a pure resistance, a voltage-divider reading of 50Ω is actually low by 1.9Ω. That is, the true resistance value is 51.9Ω when the voltage divider reads 50Ω. A reading of 100Ω is obtained with a resistor of 109.8Ω.

Surprisingly, the error is much less with reactive impedances. An impedance with a phase angle of 45° will have equal reactive and resistive components. For example, a 50Ω impedance with a positive 45° phase angle will have:

$$R = 35.36\Omega \text{ and } X_L = +35.36\Omega$$

and

$$|Z| = [(35.36)^2 + (35.36)^2]^{\frac{1}{2}} = 50\Omega$$

From *Fig. 2.17*, we see that a reading of 50Ω with a 45° reactive impedance represents a true impedance of 51.4Ω while a 100Ω reading represents a true impedance of 106.1Ω. The error when measuring a pure reactance never exceeds 0.5Ω.

Fortunately, this problem is easily fixed and the voltage divider brought back into good graces. The worst error will occur at f_{SA} where the impedance is maximum and purely resistive. Three corrective approaches are appropriate: 1) use *Fig. 2.17* to correct your reading; 2) insert a second calibrating resistor into the test setup at f_{SA}, with a value near that of R_{max}; or 3) insert a variable resistor into the test setup at f_{SA} and adjust its value to yield the same voltage reading as the driver under test. Then measure the value of this resistor with your DMM or resistance bridge to get the correct impedance.

In the previous example (2.6.1), R_{max} was measured at 72.5Ω. The simplest approach to correct the measurement is to use *Fig. 2.17*, from which we see that a reading of 72.5Ω will be about 4.3Ω low. Thus R_{max} is closer to 76.8Ω. It is a straightforward matter to recalculate r_0 and R_x and repeat the measurements for f_1 and f_2.

$$r_0 = 76.8/7.85 = 9.78, \sqrt{r_0} = 3.13 \text{ and } R_x = 24.6\Omega$$

At frequencies f_1 and f_2, the impedance magnitude is much lower and more reactive. Phase angles are commonly between 45–60°. *Figure 2.17*

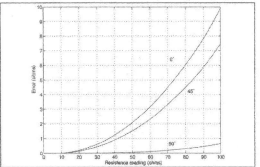

FIGURE 2.17: Voltage-divider error. (Appendix B contains a full-size version of this figure.)

shows that a 24.6Ω reactive impedance would have less than 0.3Ω error. f_1 and f_2 were remeasured at:

$$f_1 = 22.2 \text{ and } f_2 = 48.3 \text{Hz}$$

The recalculated values for the Qs are:

$$Q_{MS} = 3.88, Q_{ES} = 0.44, \text{ and } Q_{TS} = 0.40$$

Although the error in Q_{MS} is about 4% low without correction, the final value for Q_{TS} rounded off to two figures does not differ from the earlier calculation.

Figure 2.17 assumes R = 1kΩ. Because the value of R actually used was 1024Ω, the other two methods were tried to see if there was any significant additional error reduction. Using method 2, a 75Ω, 1% resistor was placed in the voltage divider and the signal-generator output adjusted to read 0.075V at 32.3Hz, again for a scale factor of 1000. With this adjustment the driver was reconnected to the divider and an R_{max} of 76.2Ω was measured.

Using method 3, a 100Ω potentiometer was inserted into the test setup and adjusted at f_{SA} to read 0.0725V (the same voltage recorded at R_{max} in the first test). The potentiometer was then removed and its resistance measured with a DC-resistance bridge, yielding 76.2Ω. From these additional measurements we see that for all practical purposes, *Fig. 2.17* can be used to correct the measurement of R_{max}.

One final point. The output voltage of typical sine-wave signal generators lies in the range of 5–7V. This, combined with the sensitivity of commonly available AC voltmeters, limits the value of R to 1kΩ, more or less. Using an amplifier to raise this voltage into the 20–30V range, R can be increased to 10kΩ. This guarantees that R is much larger than |Z| and eliminates the need for correcting the voltage divider reading. This closely approximates a true constant-current source.

(C) For drivers having a relatively low mechanical Q, the impedance curve in the region around f_{SA} will have little curvature and the corresponding maximum voltage will be quite flat. Under this condition it may be very difficult to find the exact maximum based on voltage readings.

Changes in frequency on the order of 1Hz will not perceptibly change the needle location on an analog meter. In this case Equation 2.15 may be a better estimate of f_{SA} if voice-coil inductance is known to be low.

Another approach to this problem is to use the *zero-phase property* of the impedance curve to find f_{SA}. At f_{SA} the voice-coil voltage and current will be in phase. An oscilloscope with both horizontal and vertical channel inputs can be used to generate *Lissajous patterns* from which phase can be measured. When sine-wave voltages of the same frequency and equal amplitude, but different phase, are applied simultaneously to the horizontal and vertical input channels of an oscilloscope, a closed curve is traced out, which varies from a straight line to a circle and back to a straight line as the phase angle varies from 0 to 90 to 180°. In general, Lissajous figures provide a crude phase measurement. However, the zero degree phase condition is easily discerned as a straight line.

Using the setup of *Fig. 2.18*, the free-air resonance of the 8″ driver was remeasured using the zero-phase property. The result obtained was:

$$f_{SA} = 32.2 \text{Hz},$$

which is within 0.1Hz of the value measured under the R_{max} criterion. Although we had excellent agreement between the two approaches in this example, it is often the case with very low Q_{MS} drivers that the zero phase criterion gives a more accurate result. On the other hand, drivers with large voice-coil inductance can show differences between the R_{max} and zero phase frequencies of a few tenths of a Hertz.

(D) The equation for Q_{MS} involves the *difference* of two frequencies in the denominator. In the case of low f_{SA} woofers with high mechanical Qs the difference between f_2 and f_1 can be quite small. For maximum accuracy under these conditions, f_2 and f_1 must be measured to an accuracy of 0.1Hz or better. The dial resolution on many low-cost variable-frequency sine-wave oscillators may not be adequate for reliable parameter measurements. For this reason, an audio frequency meter with 0.1Hz resolution was specified as part of the instrumentation suite. An example will help here:

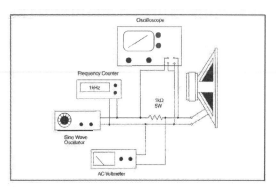

FIGURE 2.18: Test setup for measuring phase angle between voice-coil voltage and current. At resonance the oscilloscope trace will show a straight line.

EXAMPLE 2.6.2
Impedance measurements on a 10″ woofer netted the following data:

$$R_E = 6.50\Omega$$

$$R_{max} = 130\Omega \text{ at } 26.0 \text{Hz}$$

then:

$$r_0 = 130/6.5 = 20, \sqrt{r_0} = 4.47,$$

and

$$R_x = 4.47 \times 6.5 = 29.1\Omega$$

Using a frequency meter with 0.1Hz resolution, the frequencies corresponding to R_x were found to be 19.7 and 34.2Hz, respectively. From this data, Q_{MS} is calculated to be:

$$Q_{MS} = \frac{4.47 \times 26}{34.2 - 19.7} = 8.02$$

Notice that the difference between f_2 and f_1 is only 14.5Hz. Commonly available frequency meters have resolutions of 1 or 10Hz and they typically round down. With a 1Hz meter, f_1 and f_2 would read 18 and 34Hz, respectively, resulting in a calculated Q_{MS} of 7.27 for a 9% error! Calculated values Q_{ES} and Q_{TS} will also be off by 9%. This will cause roughly a 1dB error in the response of a closed-box system designed to the incorrect Q value. Clearly, a 10Hz resolution reading would be useless.

Fortunately, there is a solution to this problem. Frequency meters usually have a *period* mode. In this mode, the time it takes for the input sine wave to go through one complete cycle is measured. The frequency of a sine wave is equal to the reciprocal of its period. The 1Hz resolution meter used above

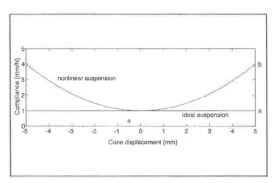

FIGURE 2.19a: Suspension compliance versus cone displacement.

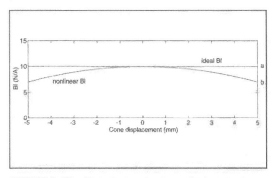

FIGURE 2.19b: B*l* product versus cone displacement.

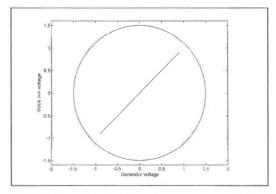

FIGURE 2.20a: Lissajous figure at resonance; small-signal condition.

also counts the period in milliseconds. The periods corresponding to f_2 and f_1, labeled T_1 and T_2, were found to be:

$$T_1 = 50.81\text{ms} = 0.05081 \text{ seconds}$$

and

$$T_2 = 29.22\text{ms} = 0.02922 \text{ seconds}$$

Then:

$$f_1 = 1/T_1 = 19.681\text{Hz and } f_2 = 1/T_2 = 34.223\text{Hz}$$

Rounded off to 0.1Hz, these are the same values measured with the 0.1Hz resolution frequency meter. The period mode is by far the best way to measure low frequencies. If your frequency meter has a period mode, use it.

(E) Accurate measurement of small-signal parameters demands a drive level small enough that all measured currents and voltages remain undistorted sine waves. Because of the high series resistance of the voltage divider, all electrical damping of the driver under test (DUT) from back EMF is lost. Under this condition, relatively low drive levels at f_{SA}, the frequency of maximum response, can produce large cone excursions, pushing the driver out of the small-signal regime and into nonlinear operation.

The two major sources of nonlinear driver response are drop-off in motor strength (B*l*) as the voice coil moves out of the gap and departure of the suspension from an ideal linear spring. Asymmetry in the magnetic fringing fields—above and below the top plate—are also a source of nonlinear behavior.

The first and third sources of nonlinearity are discussed in Chapter 1 of the *LDC*. *Figures 2.19a* and *2.19b* illustrate the nonlinear spring and B*l* problems. The compliance of an ideal spring (curve a in *Fig. 2.19a*) is constant with increasing cone excursion in either direction. Practical suspensions tend to display quadratic nonlinearity as shown in curve b. This spring response produces even-order harmonics at large excursions. Nonlinear B*l* characteristics are shown in *Fig. 2.19b*. Here, curve a represents the ideal response, namely, that B*l* is constant with increasing cone excursion. Curve b shows a typical real-world B*l* profile. Like C_{MS}, B*l* often tends to be quadratic.

When the driver under test is overdriven, the back EMF will contain harmonics of the exciting frequency. If the driver is just slightly overdriven, the second harmonic of the drive voltage will usually dominate. The same Lissajous figures used to find the zero phase point will also reveal the presence of these harmonics. As the zero phase condition is approached and the Lissajous figure begins to close, a double loop will appear in the pattern if second-order distortion is present (*Fig. 2.20*). As the drive level is reduced these loops will decrease in size and ultimately disappear. At this point the driver is back in the small-signal regime.

An example here will help. The 8″ driver in the above test was driven with an average current level of 3mA. The drive level was increased until a very slight double loop appeared in the Lissajous display indicating the presence of second harmonic. This occurred when the drive current at f_{SA} reached 10mA. The resulting measurements and T/S parameters at this drive level are presented below:

$$R_{max} = 72.5\Omega \text{ and } f_{SA} = 30.3Hz$$

then:

$$r_0 = \frac{72.5}{7.85} = 9.24$$

and

$$\sqrt{r_0} = 3.04$$

The frequencies corresponding to R_x were found to be:

$$f_1 = 20.8Hz \text{ and } f_2 = 46.2Hz$$

Checking the simplified impedance model assumptions we find:

$$f'_{SA} = (20.8 \times 46.2)^{1/2} = 31.0Hz,$$

which is within 1Hz of the measured f_{SA}. From these values, the Qs were computed as follows:

$$Q_{MS} = \frac{3.04 \times 30.3}{46.2 - 20.8} = 3.63$$

$$Q_{ES} = \frac{Q_{MS}}{9.24 - 1} = 0.44$$

$$Q_{TS} = \frac{3.63 \times 0.44}{3.63 + 0.44} = 0.39$$

In this particular example the resonant frequency and mechanical Q each dropped about 6%. The electrical Q changed only about 2%. This indicates that the increased drive level and the distortion that resulted were largely due to changes in the suspension (i.e., mechanical) parameters. From these results we can generalize to state that drive levels at f_{SA} should be limited to 10mA or less when making free-air measurements to keep the driver in the small-signal regime. One European manufacturer specifies a value of 7.5mA.

(F) When testing low-resonance, high-compliance drivers with heavy cones it is advisable to keep the driver frame vertical. Otherwise, gravity acting on the mass of the cone/voice-coil assembly may pull the cone away from its normal rest position which, in turn, moves the suspension and voice coil out of their small-signal positions. This may require mounting the driver on a small baffle with proper support to keep the driver vertical.

(G) The above test was conducted with the driver suspended approximately 5′ off the floor, the same distance from the ceiling and more than 10′ from any other interfering objects. The *LDC* indicates that slightly more accurate results can be

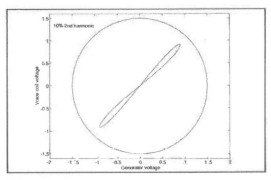

FIGURE 2.20b: Lissajous figure at resonance; nonlinear signal condition.

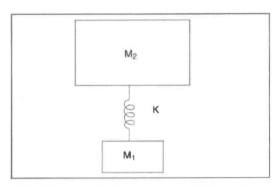

FIGURE 2.21: Double mass model for a freely suspended driver.

obtained when the driver is clamped to a test jig of some sort to prevent the driver frame from moving. Just how much more accurate are the results under this condition?

The simple spring mass mechanical model discussed in Section 2.4.1 and illustrated in *Fig. 2.3* assumes that the upper end of the spring is attached to a fixed, unmovable point. The analogy between this simple model and the driver is that the driver frame is unmovable. When the driver frame is suspended from a cable or chain, the frame is free to move.

According to Newton's third law, the motion of the cone moving relative to the driver frame will create a reaction force causing the frame to move. This situation can be approximated by the model shown in *Fig. 2.21*. It consists of two masses coupled by a spring. M_2 represents the driver frame mass, including the mass of the magnet and magnetic circuit, while M_1 is the moving mass of the cone and voice coil assembly. M_1 corresponds to the original mass, M, in *Fig. 2.3*. I leave it to the reader to devise a scheme for supporting the much larger mass, M_2.

The critical parameter here is the ratio of M_2 to M_1. The larger the mass of the frame relative to the moving mass, the less the frame will move. Let κ equal the ratio:

$$\kappa = \frac{M_2}{M_1} \qquad [2.23]$$

Then the resonant frequency and mechanical Q of the suspended driver relative to the clamped driver are given by:

$$f_{SA} = \left(\frac{\kappa+1}{\kappa}\right)f_{SAC} \qquad [2.24]$$

$$Q_{MS} = \left(\frac{\kappa}{\kappa+1}\right)Q_{MSC} \qquad [2.25]$$

where f_{SAC} and Q_{MSC} are the "clamped" values. The resonant frequency measured by hanging the driver from the ceiling is slightly higher than its clamped value. Q_{MS} is slightly lower. The 10″ driver used in a previous example weighs approximately 8 lbs (3632g). (See Appendix A for a discussion of this conversion.) The moving mass of this driver as given by the manufacturer is 49.8g. Thus κ is 73 and

$$f_{SA} = 1.014\ f_{SAC} \text{ and } Q_{MS} = 0.986\ Q_{MSC}.$$

For this driver, the measured free-hanging resonant frequency is 1.4% higher than its clamped value and the mechanical Q is 1.4% lower.

Of course, a driver suspended from a cable is not entirely free to move, so the model of *Fig. 2.21* may slightly overestimate the difference. An 8″ cast frame driver was measured under suspended and clamped conditions. The driver assembly weighed 3.5 lbs for a mass of 1589g. The cone mass was measured and found to be 22.4g. (See Section 2.7.1 for more information on this driver.) Thus:

$$\kappa = 72.3$$

and from Equation 2.24 we predict that

$$f_{SA} = 1.014\ f_{SAC}.$$

The measured values are:

$$f_{SA} = 23.4 \text{ and } f_{SAC} = 23.1$$

for a ratio of 1.013:1. These are typical results. Considering the final use of these parameters in loudspeaker design, these errors are not very significant. If desired, however, the parameters measured by suspending the driver rather than clamping it can be corrected with Equations 2.24 and 2.25. Then simply use f_{SAC} and Q_{MSC} in place of f_{SA} and Q_{MS} in your design calculations.

(H) One final caution. A dynamic loudspeaker is also a very sensitive microphone. Extraneous low-frequency noise in the test area can confuse readings. Many of you will be testing your drivers in the basement. The roar of a furnace suddenly coming to life during a test will cause the voltmeter readings to waver enough to confuse your measurements. You will have to wait until the heating cycle is finished or shut the furnace down. Just remember to turn it back on when you are done!

2.6.3 THE CONSTANT-VOLTAGE PROCEDURE

The constant-voltage procedure was first given by Thiele.[8] This procedure avoids some of the shortcomings of the voltage-divider approach. In this second approach, the speaker under test is driven from a power amplifier with a low source impedance. Under this condition back EMF damping is effective, allowing higher drive levels and thereby easing requirements on the sensitivity of the meters used in the test. It is still advisable, however, to keep drive levels at the lowest value that produces reliable results with your instruments. The price for this is the added power amp and either an AC current meter or a more complex procedure requiring two measurements at each frequency point.

The constant-voltage technique is diagrammed in *Fig. 2.22*. The following instruments are required:

1. Digital ohmmeter or DC resistance bridge
2. Calibrated variable frequency sine-wave oscillator
3. Power amplifier
4. High input impedance AC voltmeter
5. AC ammeter (optional, see text)
6. Substitute resistor, R_c
7. Frequency meter with 0.1Hz resolution
8. X,Y oscilloscope (optional)

The constant-voltage procedure uses an ohmmeter, a calibrated variable frequency sine-wave oscillator, a frequency meter, a power amplifier, an AC voltmeter, an AC milliammeter and a test resistor, R_c. Alternatively, as shown in *Fig. 2.23*, the AC voltmeter can be used to measure both the voltage across the driver terminals and, with an appropriate sampling resistor, the current flowing through the voice coil. Since we are interested only in the shape of the impedance curve, neither meter need be absolutely accurate. However, the voltmeter

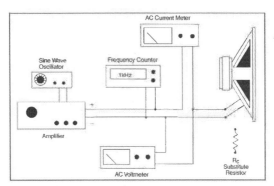

FIGURE 2.22: Test setup for constant-voltage procedure.

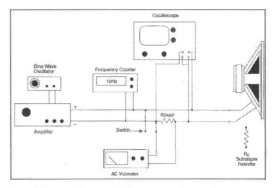

FIGURE 2.23: Constant-voltage procedure using sampling resistor and AC voltmeter in place of AC current meter. Also illustrates oscilloscope connections for measuring voice-coil phase.

must have flat frequency response to assure that the applied voltage is constant with frequency and high input impedance so as not to corrupt the current measurement.

The ammeter, if used, must have a linear frequency response so that current ratios can be computed accurately. Many digital multimeters can be used for both the resistance and AC current measurements. A power amplifier is needed because we are driving the low impedance loudspeaker directly.

If an AC voltmeter is used to measure the voice-coil current, select the smallest value sampling resistor that will produce reliable readings with your meter in order to maximize the electrical damping. A value of 1Ω or less is preferable.

The constant-voltage procedure is given below:

1. Measure the DC resistance of the voice coil with a digital multimeter (DMM) or an accurate bridge using the same procedure as that used in the voltage-divider approach.

2. Select a value for R_c close to that of R_E. R_c can be a 1% precision resistor. Alternatively, measure the value of R_c to the nearest 0.1Ω using the same instrument used to measure R_E.

3. The relative current due to R_E alone cannot be measured directly since it cannot be separated from other voice-coil impedance effects with simple instrumentation. A test resistor is used for this purpose. Set the oscillator frequency to a value near the expected value of f_{SA}. Connect R_c to the test terminals and set the amplifier output voltage somewhere in the range of 0.1–1V. The exact value is not critical. Use the lowest value that will give good results with your equipment. Call this voltage V_{test} and record its value. *This value must be held constant for all subsequent measurements.* Measure the current at the fixed voltage and label this current I_c. Calculate:

$$I_E = I_c R_c/R_E$$

4. Disconnect R_c and connect the driver to the test terminals. Suspend the driver in free-air as far away as possible from all reflecting surfaces. Keeping the voltage across the voice-coil terminals constant at V_{test}, adjust the frequency for minimum current. The frequency of the current minimum is f_{SA}. Record the minimum current and label it I_0. Calculate:

$$r_0 = I_E/I_0$$

and

$$I_x = \sqrt{r_0} \times I_E = (I_E I_0)^{1/2}$$

5. Find the frequencies f_1 and f_2 below and above f_{SA}, respectively, where the current is equal to I_x at the fixed voltage, V_{test}. Now check the validity of the simplified impedance model. Calculate:

$$f'_{SA} = (f_1 f_2)^{1/2}$$

If f'_{SA} is within 1Hz of f_{SA} or less for woofers or 1%

or less for midrange and tweeter drivers, the assumptions made in deriving the T/S parameter equations are valid. Larger differences generally imply large voice-coil inductance or nonlinear effects due to excessive drive voltage.

6. If the test of step 5 is passed, driver Qs can be calculated as follows:

$$Q_{MS} = \frac{f_{SA}\sqrt{r_0}}{f_2 - f_1}$$

$$Q_{ES} = \frac{Q_{MS}}{r_0 - 1}$$

and

$$Q_{TS} = \frac{Q_{MS}Q_{ES}}{Q_{MS} + Q_{ES}} = \frac{Q_{MS}}{r_0}$$

2.6.3.1 A CONSTANT-VOLTAGE EXAMPLE

The 8″ driver of the previous example was remeasured using the constant-voltage technique. A 6Ω, 1% resistor was used for R_c. The power amp output voltage was adjusted to obtain a current of 40mA in the 6Ω resistor. The measured voltage across this resistor was 0.24V. Then:

$$I_E = 0.04 \times (6/7.85) = 0.0306A = 30.6mA$$

Replacing the 6Ω resistor with the driver under test, the frequency was adjusted to obtain minimum current. Voltage across the driver voice coil was checked to make certain it was still 0.24V. This measurement yielded

$$I_0 = 0.0031A = 3.1mA \text{ and } f_{SA} = 32.2Hz$$

Calculating r_0 and I_x:

$$r_0 = \frac{30.6}{3.1} = 9.87$$

$$\sqrt{r_0} = 3.14$$

and

$$I_x = 9.74mA$$

The frequencies f_2 and f_1 corresponding to I_x were found to be 48.1 and 22.3Hz, respectively. From these values, the Qs were computed as follows:

$$Q_{MS} = \frac{3.14 \times 32.2}{48.1 - 22.3} = 3.92$$

$$Q_{ES} = \frac{Q_{MS}}{9.74 - 1} = 0.45$$

and

$$Q_{TS} = \frac{Q_{MS}}{r_0} = \frac{3.92}{9.74} = 0.40$$

These values agree closely with those obtained previously by the voltage-divider technique.

2.6.3.2 DRIVER PARAMETERS AS A FUNCTION OF INPUT POWER

As I stated in the beginning of this section, higher drive levels can be used with the constant-voltage procedure. Here is an example showing how driver parameters change with drive level. The driver in question is a high-quality 8″ unit with a 40mm edgewound copper voice coil, having a stated linear throw of ± 7.5mm. A 1Ω current sensing resistor was used in the tests. With this value 100mA could be passed through the voice coil at f_{SA} before exceeding the driver's linear throw limit. This is more than ten times the

TABLE 2.1
DRIVER PARAMETERS AS A FUNCTION OF
VOICE COIL VOLTAGE, V_{VC}

V_{VC}	f_{SA}	Q_{MS}	Q_{ES}	Q_{TS}
0.5	30.8	2.55	0.33	0.29
1.0	29.9	2.37	0.31	0.28
1.5	28.6	2.34	0.31	0.27
2.0	28.2	2.35	0.30	0.27
2.5	28.0	2.40	0.30	0.27
3.0	28.0	2.44	0.31	0.28
3.5	27.9	2.50	0.31	0.28
4.0	27.9	2.67	0.32	0.29
5.0	27.9	2.73	0.33	0.30

value typically obtained with the voltage-divider procedure. *Table 2.1* lists T/S parameters as a function of voice-coil drive voltage.

The impedance of this driver at f_{SA} is about 50Ω, so that voice-coil current at resonance ranged from 10mA–100mA over this set of tests. Notice that f_{SA} decreases with drive level although it does stabilize 10% lower at drive levels of 3V and above. All Qs decrease at first, but then turn around at 2.5V and above. Larger changes in driver parameters could have been obtained by pushing the driver well beyond its linear throw. I did not do this for fear of damaging the unit. The important point here is that even within its stated linear range, parameters still vary by 10%.

A smaller-value sampling resistor would have provided more damping and allowed higher drive levels, but the results would not have changed substantially. The change in driver parameters is largely a function of voice-coil excursion. With higher damping, higher drive levels are needed to get the same excursion. The major added effect of higher drive levels would be to increase voice-coil heating. This would increase R_E and lower Q_{ES}.

2.6.3.3 CAVEATS, PITFALLS, AND REMEDIES

Most of the cautions discussed in Section 2.6.2.2 regarding the voltage-divider technique also apply to the constant-voltage procedure. For example, as illustrated in *Fig. 2.23*, an oscilloscope can be used to detect the zero phase frequency and check for nonlinear response. The Lissajous comparison is made between the voltage applied to the voice-coil terminals and the current passing through the voice coil, the latter measured by the voltage across the current sampling resistor.

If an AC voltmeter is used for current measurement, keep the sampling resistor as small as possible while still getting reliable data from your meter. Since the voltage across the voice-coil terminals must be kept constant for this test, larger values of R_{samp} will cause greater variation of V_{test} with changing frequency, necessitating frequent adjustment of the signal-generator output level.

2.6.3.4 THE TRUE CONSTANT-CURRENT PROCEDURE

As discussed in Section 2.6.1, the voltage-divider procedure can be made more accurate by using a much larger resistor. Carried to its logical end, the technique will be error free if the resistor value is infinite (or at least a few megohms).

Most readers are probably familiar with constant-voltage sources. A *constant-voltage source* has a very low output impedance. As a result, output voltage changes very little with load impedance. Output current changes as required to maintain the constant output voltage. Most modern solid-state audio amplifiers have very low output impedances, often less than 0.1Ω and are thus constant-voltage sources.

A *constant-current source* (CCS) supplies a constant current to the load regardless of its impedance. The output impedance of a CCS is very high. The output *voltage* of a CCS changes as required to maintain a constant current. Thus, the output *voltage* of a CCS is proportional to its load impedance. Practical, operational-amplifier (op amp)-based AC constant-current sources have output impedances of several megohms. The following equipment is required for a constant-current measurement:

1. Sine-wave signal generator
2. High-impedance AC voltmeter
3. Electronic CCS
4. 10Ω, 1% calibration resistor (see text below)

2.6.4 THE CONSTANT-CURRENT PROCEDURE

Referring to *Fig. 2.24*, the constant-current procedure is given below:

1. Connect the calibrating resistor and AC voltmeter across the CCS output terminals. Set the signal generator-frequency near to the expected value of f_{SA}. Adjust the voltage across R_{cal} to read 0.1V. This sets the test current to 10mA. Record the signal-generator drive voltage, V_{gen}, to the CCS. This voltage must remain *constant* throughout the

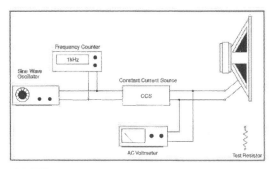

FIGURE 2.24: Test setup for constant-current procedure.

test in order to keep the CCS output current constant. This step sets the proportionality between output voltage and load impedance at 100:1. Specifically,

$$|Z| = (10\Omega/0.1V)\ V_0 = 100V_0$$

Note: Any convenient value 1% resistor can be used to calibrate the CCS. For example, if you have a 15Ω, 1% resistor, simply adjust the voltage across R_{cal} to read 0.15V.

2. Keeping the signal-generator output voltage constant at the value determined in step 2 of the constant-voltage procedure, vary the frequency to find the maximum voltage and thus the maximum impedance across the voice-coil terminals. Record the maximum impedance and label it R_{max}. Also record the frequency at R_{max} and label it f_{SA}.

3. Calculate:

$$r_0 = \frac{R_{max}}{R_E}$$

and

$$R_x = R_E \sqrt{r_0}$$

4. Find the two frequencies, f_1 and f_2, respectively, below and above f_{SA} where the impedance is equal to R_x.

5. Check the validity of the simplified impedance model by calculating:

$$f'_{SA} = (f_1 f_2)^{1/2}$$

and comparing the result to f_{SA}. If f'_{SA} is within 1Hz or less of f_{SA} for woofers, or 1% or less for midrange and tweeter drivers, the assumptions made in deriving the T/S parameter equations are valid.

6. If the test of step 5 is passed, driver Qs can be calculated as follows:

$$Q_{MS} = \frac{f_{SA}\sqrt{r_0}}{f_2 - f_1}$$

$$Q_{ES} = \frac{Q_{MS}}{r_0 - 1}$$

and

$$Q_{TS} = \frac{Q_{MS}Q_{ES}}{Q_{MS}Q_{ES}} = \frac{Q_{MS}}{r_0}$$

2.6.3.2 A CONSTANT-CURRENT EXAMPLE

Using the 8″ driver of the previous examples, I measured the following data at a drive level of 10mA:

$$R_{max} = 76.4\Omega \text{ at } f_{SA} = 32.3\text{Hz}$$

so that

$$r_0 = 9.73 \text{ and } \sqrt{r_0} = 3.12$$

then:

$$R_x = 3.13 \times 7.85 = 24.5\Omega$$

The frequencies f_1 and f_2 corresponding to R_x were found to be

$$f_1 = 22.2\text{Hz and } f_2 = 48.3\text{Hz}$$

Checking the simplified impedance model assumptions we find:

$$f'_{SA} = (22.2 \times 48.3)^{1/2} = 32.8\text{Hz},$$

which is well within 1Hz of the measured f_{SA}.

From these values, the Qs were computed as follows:

$$Q_{MS} = \frac{3.12 \times 32.3}{48.3 - 22.2} = 3.86$$

$$Q_{ES} = \frac{Q_{MS}}{9.73 - 1} = 0.44$$

and

$$Q_{TS} = \frac{3.86 \times 0.44}{3.86 + 0.44} = 0.40$$

These are the same results obtained with the corrected voltage-divider approach.

2.6.5 RELATIVE MERITS OF THE THREE PROCEDURES FOR F_{SA} AND Q

Three procedures have been presented for determining f_{SA} and the Qs from impedance data. Properly used, all three will provide accurate results. Which one is best? I have not found any one procedure uniformly superior to the others. They all have advantages and drawbacks. Some authors have discouraged use of the voltage-divider technique because, as we have seen, in its simplest application it can be error-prone. But properly applied, with the corrections presented here, it is quite accurate.

The voltage-divider technique requires the least instrumentation and is the simplest technique to apply. For the hobbyist who needs to make these measurements infrequently, it is probably the best way to go. The voltage-divider procedure also provides a simple way to measure driver impedance over the entire range of audio frequencies.

Some authors have suggested that the constant-voltage procedure is superior to the other approaches because, being driven from a power amp with low series resistance, this procedure more closely approximates the way the driver will be used. When properly accounting for drive levels to keep the driver in its linear range, however, we have seen that all three approaches give comparable answers.

The constant-voltage procedure may be less susceptible to noise, but it requires an AC ammeter, which some readers may not have. With one PC-based acoustic data acquisition and analysis system to be discussed in Chapter 7, constant-voltage measurements are easily made over a wide range of power levels.

The constant-current procedure is the easiest to use. It is especially useful for measuring the entire driver impedance curve. Unfortunately, it also requires the largest test instrument complement. If one is measuring a large number of drivers, this procedure could reduce the overall test time.

2.7 MEASURING V_{AS}

V_{AS} is the most difficult driver parameter to measure accurately. This is true for at least three reasons:

1. V_{AS} is a function of temperature, atmospheric density, and the stability of the driver suspension compliance. It varies with time and location.

2. The electroacoustic models relating V_{AS} to measured electrical parameters are only approximately correct.

3. The equations for these electroacoustic models are sensitive to errors in the measurements.

Fortunately, V_{AS} does not have to be known with extreme accuracy. A 25% variation in V_{AS} produces less than a $\pm 1dB$ variation in the passband response of the popular B_4 and QB_3 vented box alignments. A 25% variation in V_{AS} has even less effect on the passband response of popular closed box alignments ($\alpha > 3$, $Q_{TC} < 1.1$).

There are at least four procedures for measuring V_{AS} from driver impedance data. They are:

1. Free air/closed box
2. Added mass in free air
3. Open box only
4. Open box/closed box

The first two procedures, which are the most popular, will be discussed here. The last two will be covered in Chapter 3. Method 1 measures V_{AS} directly, but requires the construction of a sealed box of appropriate volume. Method 2 measures the moving mass directly, which is then combined with knowledge of f_{SA} to compute the mechanical compliance, C_{MS}, using Equation 2.4. V_{AS} is then computed using Equation 2.14. This last step requires that we know or can estimate the cone area, S_D.

2.7.1 THE FREE-AIR/CLOSED-BOX PROCEDURE

When a driver is placed in a sealed box, its resonant frequency rises. This is because the inward cone motion is resisted not only by the stiffness of its own suspension, but also by the compression of the air in the box behind it. The stiffness of the driver suspension is augmented by the stiffness of the air spring. If the total stiffness has increased, then according to Equation 2.1, the resonant frequency of the driver in the box must rise. If the stiffness of the driver/box combination has increased, then the combined compliance has

decreased. The change in resonant frequencies, along with the known box volume, are used to determine V_{AS}.

The box size used in the test must cause the *in-box resonant frequency* of the driver under test to shift upwards by 50% or more of its free-air value in order to obtain a reliable result. The in-box resonant frequency, f_{CT}, is a function of both the driver V_{AS} and the box volume, V_B. f_{CT} is given approximately by:

$$f_{CT} = \left[\left(\frac{V_{AS}}{V_B} \right) + 1 \right]^{\frac{1}{2}} f_{SA} \qquad [2.26]$$

where V_{AS} and V_B are both in the same volume units (liters, cubic inches, and so on). For example, if V_B is one-half V_{AS}, then

$$f_{CT} = (2+1)^{\frac{1}{2}} f_{SA} = \sqrt{3} \, f_{SA} = 1.73 \, f_{SA}$$

for a 73% increase in resonant frequency. If V_{AS} is known from manufacturer's data, selection of a box volume is straightforward. But if V_{AS} is known, why measure it? V_{AS} is the most difficult parameter to control in the manufacturing process. In any one sample it can easily vary from manufacturer's specifications by $\pm 30\%$.

If V_{AS} is not known, the test volumes (*Table 2.2*), first suggested in *LDC*, should cover most circumstances.

TABLE 2.2	
SUGGESTED TEST VOLUMES (IN³)	
DIAMETER	**TEST BOX VOLUME**
4–5″	220
6–7″	860
8″	1700
10″	2500
12″	3450
15″	4300

The exact test volume is not critical, but it is important to know the test volume exactly.

The free-air/closed-box procedure for determining V_{AS} is given below:

1. Measure f_{SA} and Q_{ES} using one of the procedures given in Section 2.6.
2. Mount the DUT in the test box, preferably in the orientation shown in *Fig. 2.25*. Make sure there are no air leaks around the seal between the box and the driver frame. See the discussion below for details on determining the total enclosed volume.
3. Measure the new in-box resonant frequency and electrical Q using the same procedure as that used in step 1. Label these new values f_{CT} and Q_{ECT}, respectively.
4. Compute V_{AS} as follows:

$$V_{AS} = \left[\left(\frac{f_{CT} Q_{ECT}}{f_{SA} Q_{ES}} \right) - 1 \right] V_B \qquad [2.27]$$

where V_B is the total volume behind the driver under test.

27

2.7.1.1 A V_{AS} EXAMPLE

The manufacturer of a particular cast frame 8″ woofer gives the following parameters:

$$f_{SA} = 22.1 Hz$$
$$Q_{MS} = 5.07$$
$$Q_{ES} = 0.26$$
$$V_{AS} = 149 \, ltr$$

A test box of 63.6 ltr was available. From Equation 2.26, this volume should produce an upwards frequency shift given approximately as:

$$f_{CT} = \left[\left(\frac{V_{AS}}{V_B}\right) + 1\right]^{\frac{1}{2}} f_{SA} = \left[\left(\frac{149}{63.6}\right) + 1\right]^{\frac{1}{2}} f_{SA} = 1.83 f_{SA}$$

The shift of 83% is well above the suggested 50% minimum, so this box should work out well. Free-air values of f_{SA} and Q_{ES} were first measured using the corrected voltage-divider technique. The values obtained were:

$$f_{SA} = 23.4 Hz \text{ and } Q_{ES} = 0.30$$

Next, the driver was mounted in the test box and driven to a fairly large excursion at 5Hz to check for air leaks. Following the air leak test, closed-box values for f_{CT} and Q_{ECT} were determined to be:

$$f_{CT} = 40.4 Hz \text{ and } Q_{ECT} = 0.54$$

The actual upward frequency shift obtained was only 73%. Why the difference? When placed in a box, the driver front and rear waves are isolated. Under this condition, the mass reactance loading on the cone increases over its free-air value. This accounts for some of the smaller frequency rise. Mass reactance loading of the driver cone when placed on a baffle or in an enclosure will be discussed in more detail in Chapter 3. Also, the actual driver V_{AS} differs from manufacturer's specs.

Before our final calculation of V_{AS}, the enclosed volume must be corrected for the additional volume of the mounting hole and driver cone. The mounting hole cut in 0.75″ plywood was 7.63″ in diameter for an added volume of 34.3 in^3 or 0.56 ltr. The calculated cone volume was 0.34 ltr for a total volume of 64.5 ltr. The corrections for hole and cone volume amount to only 1.4% and may well be insignificant, but it is instructive to go through the calculation anyway. With the corrected volume we are now in a position to compute V_{AS}.

$$V_{AS} = \left[\left(\frac{40.4 \times 0.54}{23.4 \times 0.30}\right) - 1\right] \times 64.5 = 136 \, ltr$$

The measured value of f_{SA} is about 5% higher than manufacturer's specification. Since f_{SA} varies inversely as the square root of mechanical compliance and V_{AS} is directly proportional to C_{MS}, we would expect V_{AS} to be about 10% smaller than manufacturer's specification. And this is what was obtained.

2.7.1.2 DISCUSSION

(A) V_B must include all the volume behind the driver cone. This includes the volume of the mounting hole and the effect of the driver volume. *Figure 2.25* shows the driver under test mounted from the outside of the test box. If a good seal can be obtained between the driver frame and the box, this is the preferred mounting arrangement. With this arrangement, the volume added by the driver's truncated cone is easily determined. If the driver is mounted in the conventional manner, the volume displaced by the cone plus the magnet assembly must be determined and subtracted from the box volume. This is a more difficult task. Referring to *Fig. 2.26*, the volume of a truncated cone may be computed with the following steps:

1. Measure D_1, D_2, and h;
2. Draw the trapezoid a,a, D_1,D_2 to scale on quadruled paper and extend the "a" lines until they meet;
3. Determine h_1 and h_2 from the scaled drawing;
4. Compute the truncated cone volume as follows:

$$V_{cone} = \left(\frac{\pi}{12}\right)\left(D_1^2 h_1 - D_2^2 h_2\right) = 0.262\left(D_1^2 h_1 - D_2^2 h_2\right)$$

(B) The test box must be well sealed. Large air leaks may cause incorrect results. Once the driver is mounted, the box can be checked for air leaks by exciting the driver with a low-frequency sine wave, preferably in the range of 5–10Hz. Apply enough power to get substantial, but not excessive (i.e., damaging), cone excursion. Place your ear near areas of suspected leaks. Any leak will be heard as a

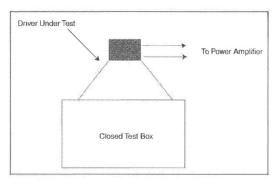

FIGURE 2.25: Test setup for measuring V_{AS} using the closed-box procedure.

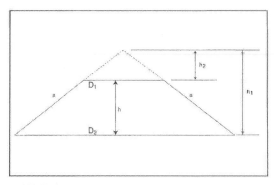

FIGURE 2.26: Diagram for computing volume of truncated cone.

chuffing noise or felt as a pulsating air stream. The leak is easily plugged with Mortite™, Blue Tac™, or any other non-hardening sealer.

(C) The walls of the test box should be reasonably stiff, but there is no need to go overboard. Energy absorption due to box-wall flexing is reflected only in Q_{MCT}—the driver in-box mechanical Q—and does not affect the measured value of Q_{ECT}. Do not, however, place any absorbing lining in the box, as this will change the effective value of V_B. (See Chapter 3 for a discussion of the effect of box lining and stuffing on V_B.)

2.7.2 THE ADDED-MASS TECHNIQUE

In the free-air/closed-box procedure, resonant frequency is *raised* by placing the driver in a sealed box. With the added-mass procedure, driver resonant frequency is *lowered* by adding mass to the cone. In this method, the free-air resonant frequency of the driver is measured first. Then a known mass is added to the cone and the new, lower, free-air frequency is measured. This data is sufficient to calculate the driver moving mass, M_{MS}. M_{MS} and f_{SA} are used next to compute driver mechanical compliance, C_{MS}, using Equation 2.4. Finally, C_{MS} and S_D are used to compute V_{AS} with Equation 2.14.

For this method to work well, you should try to get a downward frequency shift of 25% or more. For large woofers this may mean adding 30–60g to the cone. The added mass should be placed symmetrically about the cone apex. Use material such as Mortite or non-hardening putty. Blue Tac non-hardening putty, available in most hardware stores, works especially well.

Small ceramic disk magnets also work well as long as the driver frame is nonmagnetic. Placing a magnetic disk on each side of the cone with poles attracting will clamp the disk pair to the cone. The added mass must be weighed accurately, preferably to within 0.1g. As with the free-air tests, the driver should be vertical to prevent preloading the suspension. This is especially important because of the extra weight of the added mass. Finally, if S_D is not given by the manufacturer, you can calculate it to a good approximation by using a diameter which includes one-half of the surround, since a portion of the surround moves with the cone and contributes to the effective piston area.

The added-mass procedure for determining V_{AS} is given below:

1. Measure f_{SA} using one of the procedures given in Section 2.6.
2. Add a known mass to the cone. If the cone mass is known from manufacturer's data, add a mass equal to at least 60% of the cone mass. Measure the added mass to 0.1g accuracy.
3. Measure the new free-air resonant frequency using the same procedure as in step 1. Label this frequency f_{SAM}. If this frequency is not at least 25% lower than f_{SA}, add additional mass and repeat step 3.
4. Compute moving mass using the following formula:

$$M_{MS} = \frac{M_{ADD}}{\left(\dfrac{f_{SA}}{f_{SAM}}\right)^2 - 1} \qquad [2.28]$$

where M_{ADD} is the added mass.

5. Compute driver mechanical compliance using:

$$C_{MS} = \frac{1}{4\pi^2 f_{SA}^2 M_{MS}} = \frac{0.0253}{f_{SA}^2 M_{MS}} \qquad [2.29]$$

6. Finally, compute V_{AS} using Equation 2.14:

$$V_{AS} = \rho c^2 C_{MS} S_D^2 = 1.4 \times 10^5 C_{MS} S_D^2, \qquad [2.30]$$

where S_D is in m^2 and standard atmospheric conditions have been assumed in evaluating the ρc^2 term.

2.7.2.1 AN ADDED MASS EXAMPLE

Let's determine V_{AS} for the 8″ woofer in example 2.7.1.1, using the added-mass method. The free-air resonant frequency is already known from the previous example. The moving mass and cone area of this driver are specified at 22.2g and 0.0222m², respectively, by the manufacturer. Three blobs of Blue Tac totaling 23.8g were placed symmetrically about the cone apex and the new free-air resonant frequency was measured yielding:

$$f_{SAM} = 16.3 \text{Hz}$$

Then using Equations 2.28–2.30 we obtain:

$$M_{MS} = \frac{23.8}{\left(\dfrac{23.4}{16.3}\right)^2 - 1} = 22.4g = .0224 \text{kg}$$

$$C_{MS} = \frac{0.0253}{(23.4)^2 \times 0.0224} = 0.00206 \text{m / N}$$

$$V_{AS} = 1.4 \times 10^5 \times 0.00206 \times 0.0222^2 = 142 \text{ ltr}$$

This result is within 4% of the result obtained using the free-air/closed-box method. This level of agreement is excellent for V_{AS} measurements.

2.7.2.2 DISCUSSION OF ADDED-MASS TECHNIQUE

(A) The term ρc^2 changes by about 0.4% as the temperature drops from 68–32°F (20°C–0°C) at sea-level.

(B) The act of attaching a test mass to the cone will often shift the driver compliance due to suspension hysteresis or creep. This happens because the cone position is shifted slightly when applying the test mass. Exciting the driver at resonance for a minute or so with the added mass in place will usually reset this suspension shift. It is possible to add too much mass to the cone, throwing it out of alignment or preloading it sufficiently to move the suspension out of its linear range.

Do not add a mass larger than the cone mass

itself. Breaking the added mass into three or four equal elements and placing them symmetrically about the cone apex will help to prevent these problems. Ceramic disk magnets provide a very convenient way to add cone mass. With stamped steel frame drivers, however, the magnets may pull the cone toward the frame producing a compliance shift like that discussed above.

2.7.3 COMPARISON OF ADDED-MASS AND CLOSED-BOX PROCEDURES FOR DETERMINING V_{AS}

Obviously, the added-mass technique is simpler to implement. You don't have to build a box. Unlike the closed-box method, the added-mass technique is not sensitive to leakage errors. Errors may be incurred, however, due to the uncertainty in S_D and uncompensated compliance shifts. The closed-box method measures V_{AS} directly. It is not subject to errors in S_D and does not normally experience compliance shifts. However, leakage errors can occur even in a properly sealed box due to leakage through the driver surround or dust cap.

The added-mass technique measures cone mass, M_{MS} and compliance, C_{MS}, directly with little error. The closed-box method measures V_{AS} with little error. Using data from both techniques you can solve for S_D. The appropriate equation is:

$$S_D = 10\left(\frac{2\pi M_{MS} V_{AS} f_{SA}^2}{\rho c^2}\right)^{1/2} = 0.168 f_{SA}\left(M_{MS} V_{AS}\right)^{1/2},$$

where M_{MS} is in grams and V_{AS} is in liters. Using the data from our examples:

$$S_D = 0.168 \times 23.4 \times (22.4 \times 136)^{1/2} = 217 cm^2 = 0.0217 m^2$$

This result is within 2.3% of manufacturer's specification.

2.8 THE REST OF THE T/S PARAMETERS

Some of the additional T/S parameters can be measured directly with additional instrumentation, but it is now common practice throughout the industry to calculate these parameters from the complete electro-mechano-acoustic model of the driver. The parameters we have learned how to measure so far, f_{SA}, V_{AS}, and all three Qs, are sufficient to calculate all of the other T/S parameters of interest.

2.8.1 CALCULATING CONE MASS

Cone moving mass is determined directly via the added-mass technique. However, if you determined V_{AS} using the closed-box procedure, you may wish to calculate the cone mass. This can be done in two steps:

1) Use V_{AS} to calculate driver mechanical compliance, C_{MS};
2) Use f_{SA} and C_{MS} to compute M_{MS}.

Solving Equation 2.29 for C_{MS} and 2.28 for M_{MS} we get:

$$C_{MS} = \frac{V_{AS}}{\rho c^2 S_D^2} = 7.16 \times 10^{-6}\left(\frac{V_{AS}}{S_D^2}\right) \quad [2.31]$$

and

$$M_{MS} = \frac{0.0253}{f_{SA}^2 C_{MS}} \quad [2.32]$$

Let's continue with our 8″ driver example from Section 2.7.1:

$$C_{MS} = 7.16 \times 10^{-6} (0.136/0.0222^2) = 0.00198 \text{ N/m} = 1.98 \text{ mN/m}$$

and

$$M_{MS} = 0.0253/(0.00198 \times 23.4^2) = 0.0233 kg = 22.3 g$$

This value is in good agreement with the manufacturer's specification of 22.2g.

2.8.2 CALCULATING DRIVER Bl PRODUCT

The magnetic field intensity, B, can be measured only by removing the driver cone assembly and inserting a magnetic sensor such as a Hall-effect sensor directly into the voice-coil gap. Fortunately, we don't have to measure B. Driver performance is a function of the Bl product. *LDC*, Section 8.43, gives one technique for measuring Bl directly; however, it is also easily and more commonly calculated from the T/S model. The appropriate equation is:

$$Bl = \left(\frac{2\pi f_S R_E M_{MS}}{Q_{ES}}\right)^{1/2} = 2.51\left(\frac{f_S R_E M_{MS}}{Q_{ES}}\right)^{1/2} \quad [2.33]$$

Bl has the units of newtons/ampere or tesla-meters. Notice that we have used f_S in Equation 2.33, rather than f_{SA}. When mounted on infinite baffle or in an enclosure, driver front- and back-waves are isolated, causing additional acoustic mass loading on the cone. This added mass lowers the resonant frequency slightly and changes the driver Qs relative to their free-air values. This has been mentioned before and will be discussed fully in Chapter 3. For the moment we will use free-air parameters in the following examples, recognizing that our results will be slightly in error.

For the cast frame driver we have been using as an example, R_E is 7.55Ω. Then:

$$Bl = 2.51\left(\frac{23.4 \times 7.55 \times 0.0223}{0.30}\right)^{1/2} = 9.1 N/A$$

The manufacturer lists Bl at 9.3N/A indicating that this particular sample is very close to manufacturer's specs.

2.8.3 EFFICIENCY AND SENSITIVITY

Perhaps the most important of the T/S parameters to be calculated is driver *efficiency*. This parameter tells us how much acoustic power and sound-pressure level a driver can generate for each electrical watt of input power. It is also useful in matching drivers in multiway loudspeaker systems. Woofer, midrange, and tweeter drivers should have comparable sensitivities in order to produce a well-balanced frequency response. Small[11] has shown that the expression for the *half-plane, mid-band* efficiency can be written in terms of the driver resonant frequency, f_S, driver electrical Q_{ES}, and V_{AS}. The appropriate equation is:

$$\eta_0 = \frac{4\pi^2 f_S^3 V_{AS}}{c^3 Q_{ES}} \qquad [2.34]$$

where η_0 is the loudspeaker efficiency (acoustic watts out per electrical watt in) and the other terms have been defined previously. The expression "half-plane, mid-band" means that this is the efficiency attained by the driver when mounted in the center of a very large wall (i.e., an infinite baffle) and operating at frequencies well above resonance, but still in the piston range.

The full-space efficiency is 3dB less. As with Equation 2.33, we see f_S rather than f_{SA} in Equation 2.34. The comments given in connection with Equation 2.34 apply here also. Again, we will use free-air parameters in the following examples, recognizing that our results will be slightly in error.

Equation 2.35 can be written in more convenient terms:

$$\eta_0 = 9.6 \times 10^{-10} \frac{f_S^3 V_{AS}}{Q_{ES}} \qquad [2.35]$$

where V_{AS} is in liters. Notice that efficiency is proportional to the cube of f_S and thus very sensitive to this parameter. Equation 2.35 shows why it is so difficult to get low frequency extension and efficiency at the same time. For example, reducing f_S by a factor of two, in order to pick up an octave of bass extension while maintaining a fixed Q_{ES} and V_{AS}, would lower efficiency by a factor of eight. Alternatively, to maintain the same efficiency with a fixed Q_{ES}, V_{AS} would have to be increased by a factor of eight. This would translate directly into an eight-fold-increase enclosure volume.

For typical listening environments there is a direct relationship between radiated acoustic power and SPL. This allows us to express power sensitivity in terms of efficiency. Power sensitivity is related to efficiency by:

$$S_p = 112.2 + 10\log (\eta_0) \text{ dB SPL/1W/1m} \quad [2.36]$$

Power sensitivity specifies the sound-pressuure level produced by the driver when mounted on a large baffle at a distance of 1m in response to an electrical input of 1W.

Let's calculate the efficiency and power sensitivity of the 8″ driver example of Section 2.7.1. Although not strictly correct, we will use the free-air driver parameters in this calculation. The error will be small. Recall the measured parameters for this driver are:

$$f_{SA} = 23.4, Q_{ES} = 0.30, \text{ and } V_{AS} = 136 \text{ ltr}$$
(closed-box method)

Then:

$$\eta_0 = 9.6 \times 10^{-10} \left[\frac{(23.4)^3 \times 136}{0.30} \right] = 0.0056 = 0.56\%$$

and

$$S_p = 112.2 + 10\log (0.0056) = 112.2 - 22.5 = 89.7 \text{dB/1W/1m}$$

For this driver, an electrical input of 1W will produce an output of 5.6 acoustic milliwatts and a sound-pressure level of 89.7dB at 1m in front of the driver. Notice that the logarithm of a number less than one is negative.

Efficiency can also be expressed in terms of the fundamental electromechanical driver parameters. This expression for the half-plane, mid-band efficiency is:

$$\eta_0 = \frac{\rho (Bl)^2 S_D^2}{2\pi R_E M_{MS}^2} = 5.44 \times 10^{-4} \frac{(Bl)^2 S_D^2}{R_E M_{MS}^2} \quad [2.37]$$

This expression for efficiency is entirely equivalent to that of Equation 2.35. In fact, Equation 2.37 was known long before the work of Thiele and Small. It was Small who derived Equation 2.35 from 2.37, giving us a different and very useful insight into the driver efficiency issue. Small showed us the interrelationship between efficiency, low-frequency extension, and enclosure size. In particular, Equation 2.35 demonstrates that extended low-frequency response and high efficiency are incompatible with compact size.

Unlike Equation 2.35, Equation 2.37 has some intuitive appeal. For example, we would expect a stronger magnetic field to produce more output. We would also expect a heavier cone to be less efficient. Similarly, all things being equal, a larger cone area should radiate more power. Equation 2.37 is interesting because it highlights an underlying assumption in the efficiency calculation which is not obvious in Equation 2.35. The efficiency calculation assumes that the driver impedance is resistive and equal to R_E. Then the input power is simply:

$$P_E = V_g^2 / R_E$$

This approach was taken because, as we have seen, the actual driver impedance varies widely with frequency, and calculation of the input power at any one frequency is problematical. η_0 is actually a pseudo-efficiency.

Let's recalculate η_0 using Equation 2.37:

$$\eta_0 = 5.44 \times 10^{-4} (9.1^2 \times 0.0217^2)/(7.55 \times 0.0223^2) = 0.0056 = 0.56\%,$$

which is the same result we obtained with Equation 2.35.

31

The assumption of a constant resistive impedance in the efficiecny calculation means that driver frequency response tests which are nominally made with constant input power are actually made with a constant input voltage. Modern loudspeaker design usually assumes the system is driven from a low impedance voltage source. It is becoming common practice to specify driver performance in terms of a *voltage* sensitivity, S_0. S_0 has the units of dB SPL/2.83V/1m. 2.83V represents the voltage that will produce one watt of power dissipation in an 8Ω resistor. To the extent that R_E differs from 8Ω, S_P and S_0 will differ. The expression for S_0 is:

$$S_0 = 20\log_{10}\left(\frac{Bl \times S_D}{R_E M_{MS}}\right) + 88.5\,\text{dB SPL}\,/\,2.83\text{V}\,/\,1\text{m} \quad [2.38]$$

Again, using our 8″ driver example:

$$S_0 = 20\log_{10}\left(\frac{9.1 \times 0.0217}{7.55 \times 0.0223}\right) + 88.5 = 89.9\,\text{dB SPL}\,/\,2.83\text{V}\,/\,1\text{m}$$

This result differs slightly from that obtained in Equation 2.37 because R_E is slightly less than 8Ω.

2.8.4 MEASURING VOICE-COIL INDUCTANCE

Voice-coil inductance is another difficult parameter to measure, primarily because it is not constant, but varies with frequency. Furthermore, it is not a very interesting or useful parameter. From a speaker system design standpoint, the driver impedance curve (both magnitude and phase) is what is needed. The manner in which voice-coil inductance varies to produce the driver impedance curve is not important because the impedance curve can be measured directly. Having said that, manufacturers continue to list voice-coil inductance, usually at 1kHz, so we will treat it here.

Let's revisit *Fig 2.2*. This is the impedance plot for the 8″ cast frame driver used in several previous examples. The impedance magnitude for this driver is rising very sharply as the frequency approaches 20kHz. It

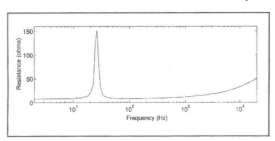

FIGURE 2.27a: Resistive part of driver impedance.

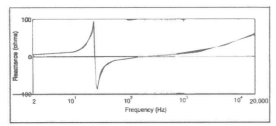

FIGURE 2.27b: Reactive part of driver impedance.

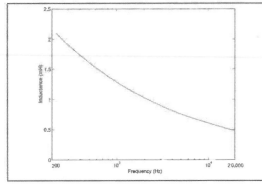

FIGURE 2.28: Voice-coil inductance versus frequency.

seems reasonable to assume that the rise is dominated by voice-coil inductance. If that were true, however, we would expect the impedance phase angle to approach 90°. But the phase angle plot shows a phase angle of only 52.3° at 20kHz. What's happening here?

Figure 2.27 gives us a very different picture of the driver impedance curve. This figure shows the real or resistive and imaginary or reactive components of driver impedance versus frequency. The real part starts out at a value slightly above R_E at 2Hz and reaches its first maximum at f_{SA} as one would expect. But beyond a few hundred hertz, it begins to rise again reaching a value of 46.1Ω at 20kHz.

The reactive component starts out positive (inductive) below f_{SA}, goes through zero at f_{SA}, goes negative (capacitive) above f_{SA}, and then passes through zero again at 208Hz before continuing its climb back into the positive region where it stays. The reactance at 20kHz is $+59.8\Omega$. The second purely resistive impedance point at 204.7Hz coincides with the zero phase frequency shown in *Fig. 2.2* as one would expect.

We see that the driver impedance rise at higher frequencies has both strong resistive and inductive components. The rising resistive component is caused by eddy currents generated in the pole piece by the AC current passing through the voice coil. These currents circulating in the pole piece generate heat loss which reflects back into the voice-coil impedance as a resistive loss. These losses increase with frequency giving rise to the growing resistive component in the driver impedance. The voice-coil inductance actually decreases with frequency due to the falling permeability of the pole piece at higher frequencies. A plot of voice-coil inductance for this driver is shown in *Fig. 2.28*.

The complex process going on here makes measurement of voice-coil inductance problematic. Voice-coil inductance has typically been measured by blocking the cone, preventing it from moving, and measuring the voice-coil inductance with an AC inductance bridge. The value is commonly specified at 1kHz. If the cone is not blocked, the electromotive impedance represented by the parallel RLC circuit in *Fig. 2.11* will add to the measured impedance and make separation of the voice-coil inductive component very difficult.

Most readers will not have the special jig required to block cone motion so that alternate, approximate techniques for getting a handle on voice-coil inductance are required. One convenient frequency at which the inductance can be determined is the second zero phase point at 204.7Hz in our example. This frequency can be determined with a Lissajous phase measurement. Because this frequency is on the order of ten times f_{SA} or more, the zero phase condition can be modeled as a series resonance between the capacitor, C_{MES}, and the voice-coil inductance, L_{VC}. The reactance of the two elements are equal at this frequency and thus cancel, giving a purely resistive impedance. We have:

$$1/(2\pi f_0 C_{MES}) = 2\pi f_0 L_{VC},$$

where f_0 is the zero phase frequency. Solving for

L_{VC} we get:

$$L_{VC} = \frac{1}{4\pi^2 f_0^2 C_{MES}} = \frac{0.0253}{f_0^2 C_{MES}} \qquad [2.39]$$

C_{MES} is computed using Equation 2.10 for the cast frame 8″ driver previouly measured.

$$C_{MES} = M_{MS}/(Bl)^2 = 0.0224/(9.1)^2 = 269.3\mu F$$

Then:

$$L_{VC} = 0.0253/[(204.7)^2 \times 0.0002693] = 2.24mH$$

In order to determine voice-coil inductance with reasonable accuracy at frequencies above the second zero phase point, both the magnitude and phase of the driver impedance are required. It can be computed using the following steps:

1. At each frequency of interest, compute the reactive component of driver impedance:

$$X = |Z| \sin \theta,$$

where $|Z|$ is the impedance magnitude and θ is the impedance phase angle.

2. At each frequency, f, for which L_{VC} is desired, correct the reactance determined in step 1 for the effect of C_{MES} using

$$X_L = X + 1/(2\pi f\, C_{MES}) = X + 0.1592/(f\, C_{MES})$$

3. Compute L_{VC} using

$$L_{VC} = 0.1592\, X_I/f$$

Let's compute L_{VC} for our 8″ cast frame driver at 1kHz. From $Fig.\ 2.2$, the impedance magnitude and phase are:

$$Z = 14.57\Omega \text{ and } \theta = 36.2°$$

Then the reactive impedance component is:

$$X = 14.57 \times \sin(36.2) = 8.61\Omega,$$

and the inductive reactance is:

$$X_L = 8.61 + 0.1592/(1000 \times 0.0002693) = 8.61 + 0.59 = 9.2\Omega,$$

and finally, the voice-coil inductance is computed to be:

$$L_{VC} = 0.1592 \times 9.2/1000 = 1.51mH$$

Notice that the effective voice-coil inductance has dropped from 2.24mH to 1.51mH in going from 204Hz to 1kHz. Repeating the above calculations at 20kHz, we find that the effective voice-coil inductance is only 0.47mH. $Figure\ 2.28$ is a plot of the voice-coil inductance over frequency for the driver in this example. In Chapter 7 I will introduce an approximate model for voice impedance that can be used with PC-based acoustic measurement systems to obtain very accurate measurements of T/S parameters.

2.9 T/S PARAMETERS FOR MIDRANGE AND TWEETER DRIVERS

The T/S parameter measurement examples I have given so far have been limited low-frequency (woofer) drivers. This was so because these parameters are most useful for the design of low-frequency loudspeaker enclosures. T/S parameters for midrange and tweeter drivers can, however, be determined using any of the techniques already described. The procedures are identical with one exception. Some midrange units and virtually all tweeters come with enclosed rear volumes. For these drivers, V_{AS} cannot be measured by the free-air/closed-box method. Compliance can be measured using the added-mass method, but since the driver comes with its own enclosure, this result is not too useful.

Many midrange and tweeter drivers have a ferrofluid damping compound in the gap between the top plate and pole piece to damp cone motion and provide better heat transfer from the voice coil to the top plate, and ultimately ambient air. If the damping is very heavy, the peak of the impedance curve at resonance may flatten to the point where it is impossible to determine T/S parameters by methods given in this chapter. An example may help here.

$Figure\ 2.29$ shows impedance curves for two tweeters. The first is a popular concave metal-dome tweeter without ferrofluid damping. The second is a soft-dome unit with heavy damping. The metal-dome unit has a well-shaped impedance curve with a pronounced resonance peak

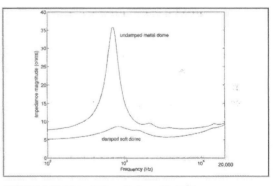

FIGURE 2.29: Tweeter impedance comparison.

at 700Hz and good symmetry about the resonant frequency point. The impedance curve for the damped soft dome displays a highly flattened resonance with poor symmetry about its maximum value at 800Hz. Clearly, the soft-dome tweeter will fail the symmetry test given by Equation 2.15. The total Q of the soft dome tweeter can be estimated by fitting a model to the acoustic frequency response of the tweeter.

2.10 SUMMARY

This chapter has emphasized the determination of T/S $small\text{-}signal$ parameters using relatively simple measurements of the driver impedance curve. This approach is widely used by driver manufacturers and is believed to be the most easily accessible to the reader, since it can be accomplished with relatively simple instrumentation. We will cover several PC-based applications in Chapter 7 that also make use of the impedance curve. These programs and their accompanying instrumentation automate the procedures discussed in this chapter and may provide more accuracy.

There are other ways to get the T/S parameters. Some involve direct electromechanical measurement of parameters such as Bl and C_{MS}. In another approach, a laser velocimeter is used to measure cone velocity and find f_{SA} at the frequency of maximum cone velocity. Others measure the parameters over the entire range of cone excursion, thus characterizing the nonlinear behavior of the driver.

Although not covered here, there is growing interest and capability within the industry to characterize nonlinear, large-signal, loudspeaker performance. Nonlinear models are gaining in importance as sophisticated software to analyze loudspeaker performance under large-signal conditions becomes available.

Reference 13 presents one of the simplest approaches to obtaining large-signal models for loudspeaker behavior. The authors apply a DC current to the voice coil to offset its displacement from its rest position and then apply a small AC signal to measure the impedance and T/S parameters at this offset location. In this manner it is possible to develop a model valid over the entire excursion capability of a driver. The model can be used to predict distortion and compression at high power levels. A more sophisticated, state-of-the-art facility using the most advanced techniques currently available for determining large-signal T/S parameters is described in Reference 14.

REFERENCES

1. Dickason, Vance, *Loudspeaker Design Cookbook*, fifth edition, Audio Amateur Press, 1995.

2. Kellogg, E.W., and Rice, C.W., Journal of the American Institute of Electrical Engineers, Vol. 44, p. 982, 1925.

3. McLachlan, N.W., *Loudspeaker, Theory, Performance, Testing and Design*, Dover Publications, Inc., 1960.

4. Olson, H.F., *Elements of Acoustical Engineering*, D. Van Nostrand Company, Inc., 1940 and 1947.

5. Olson, H.F., *Dynamical Analogies*, D. Van Nostrand Company, Inc., 1943.

6. Beranek, Leo, *Acoustics*, McGraw-Hill, 1954, and Acoustical Society of America, 1986.

7. Novak, J.F., "Performance of Enclosures for Low-Resonance High-Compliance Loudspeakers," IRE Transactions on Audio, January/February 1959.

8. Thiele, A.N., "Loudspeakers in Vented Boxes," Proceedings of the IRE, Australia, August 1961, reprinted in *JAES*, May/June, 1971.

9. Benson, J.E., "Theory and Design of Loudspeaker Enclosures, Parts 1, 2, and 3," AWA Technical Review, Vol. 14, No. 1, 1968, Vol. 14. No. 3, 1971, and Vol. 14, No. 4, 1972.

10. Small, R. "Direct Radiator Loudspeaker System Analysis," *JAES*, June 1972.

11. Small, R., "Closed Box Speaker Systems, Parts 1 and 2," *JAES*, September 1972 and January/February 1973.

12. Small, R. "Vented-Box Loudspeaker Systems, Parts 1, 2, 3, and 4, *JAES*, June 1973, July/August 1973, September 1973, and October 1973.

13. Scott, J., and Kelly, J., "New Method of Characterizing Driver Linearity," *JAES*, April 1996.

14. Clark, D., "Precision Measurement of Loudspeaker Parameters," *JAES* 99th Convention, October 1996, New York, preprint no. 4081.

15. Williamson, M., "Measuring Loudspeaker Resonance," *Speaker Builder* 3/97.

APPENDIX A
CONVERSION FACTORS

Conversion factors are listed below for the units commonly used in loudspeaker testing.

Length conversions:

To convert	Into	Multiply by
mm	cm	0.1
	m	0.001
	in	0.03937
cm	mm	10
	m	.01
	in	0.3937
m	cm	100
	in	39.37
	ft	3.281
in	mm	25.4
	cm	2.54
	m	0.0254
ft	m	0.305
	cm	30.5

Area conversions:

To convert	Into	Multiply by
cm^2	m^2	10^{-4}
	in^2	0.155
	ft^2	1.0764×10^{-3}
m^2	cm^2	10^4
	in^2	1550
	ft^2	10.764
in^2	cm^2	6.452
	m^2	6.452×10^{-4}
	ft^2	0.006945
ft^2	in^2	144
	cm^2	929.1
	m^2	0.09291

Volume conversions:

To convert	Into	Multiply by
cm^3	m^3	10^{-6}
	in^3	0.01602
	ft^3	3.531×10^{-5}
	ltr	0.001
m^3	cm^3	10^6
	in^3	6.103×10^4
	ft^3	35.32
	ltr	1000
liters	cm^3	1000
	m^3	0.001
	in^3	61.03
	ft^3	0.03532
in^3	cm^3	16.387
	m^3	1.639×10^{-5}
	ltr	0.01638
	ft^3	5.787×10^{-4}
ft^3	in^3	1728
	cm^3	2.832×10^4
	m^3	0.02832
	ltr	28.32

Force/Mass and force/force conversions:

Force/Mass conversions require some explanation. The Imperial or English system of units commonly uses the pound which is a unit of force or weight. The SI system uses the kilogram which is a unit of mass. What is the difference? On earth, as a result of gravity, you have both weight and mass. In outer space you are weightless, but you still have mass. Mass is a measure of inertia, a quality of matter which requires that a force must be exerted on a body to accelerate it. Mass/force conversions are related by Newton's second law which states that:

where:

$$a = F/M \qquad [A.1]$$

a = body acceleration with respect to inertial space

F = applied force

Newton's second law states that the acceleration of a body is directly proportional to the applied force and inversely proportional to the body's mass. Thus mass is that property of a body which resists a change in motion. The law applies whether you are on earth or in space. The law is more commonly stated in the form:

$$F = Ma \qquad [A.2]$$

The force of gravitational attraction which the earth exerts on a body is called the weight of the body. Ignoring centripetal acceleration,

$$W = Mg \qquad [A.3]$$

where:

W = a body's weight in newtons or pounds
M = a body's mass in kilograms or slugs
g = acceleration due to gravity in m/sec^2 or ft/sec^2

The average value of earth's gravity is:

$$g = 32.174 \ ft/sec^2 = 9.807 m/sec^2$$

Force/mass conversions can be likened to conversions from apples to oranges. When we say that one kilogram equals 2.2 lb., we are really saying that a mass of one kilogram weighs 2.2 lb. on the earth's surface. The actual conversion is a two step process. First, the kilogram is converted to its weight in newtons using Newton's second law, then the weight in newtons is converted to weight in pounds.

The unit of mass in the Imperial or English system is the slug. From Equation A.3 we see that one slug weighs 32.174 lb. on earth. The slug is not used in loudspeaker analysis or design. Conversions for this unit will not be given.

To convert	Into	Multiply by
gm	kg	0.001
	newtons	9.807×10^{-3}
	oz(weight)	0.03527
	lb(weight)	2.205×10^{-3}
kg	gm	1000
	newtons	9.807
	lb(weight)	2.205
	oz	35.27
oz(weight)	gm	28.35
	kg	3.527×10^{-5}
	lb(weight)	0.0625
lb(force)	newtons	4.448
	gm	453.6
	kg	0.4536

APPENDIX B

A full-size version of *Fig. 2.17* is given here for use in correcting impedance measurements made with the voltage-divider technique assuming a 10Ω calibration resistor is used with R = 1000Ω. To use this graph, find the measured value of impedance on the horizontal axis. From this point extend a vertical line until it intersects the 0°, 45°, or 90° line, depending on how reactive you believe the measured impedance to be. From the point of intersection of the vertical line with the chosen curve, extend a horizontal line to the left until it intersects the vertical axis of the graph. Read off the voltage divider error on the vertical axis. The error is *added* to the measured value to get the true impedance magnitude.

As a general rule when determining T/S parameters, use the 0° curve to correct the measured value at f_{SA} and the 45° curve to correct the measured values of f_1 and f_2. For example, a value of 80Ω is measured at f_{SA}. The 80Ω line intersects the 0° curve at 6Ω. Thus the true value is $80 + 6$ or 86Ω. For measured values above 1000Ω it is best to recalibrate the voltage divider with a higher value resistor as explained in Chapter 2.

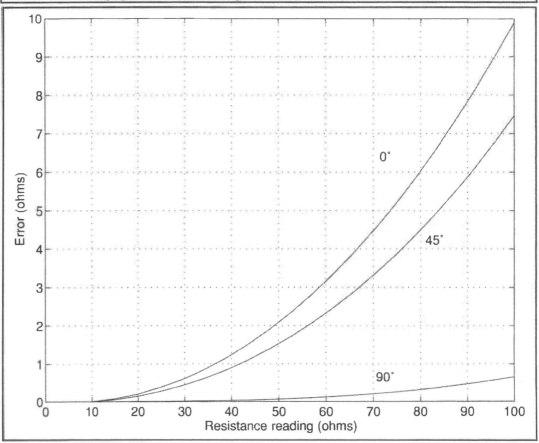

FIGURE 2.17: Voltage-divider error.

CHAPTER 3

LOW-FREQUENCY SYSTEM ELECTRICAL IMPEDANCE TESTS

3.0 INTRODUCTION

In this chapter we will continue to learn about loudspeaker low-frequency performance from the impedance curve. Now, however, we will examine the impedance characteristics of drivers mounted in enclosures; particularly, the impedance curves for closed-box and vented-box systems—in order to determine parameters, such as system-resonant frequency or frequencies, enclosure damping or Q, system efficiency, and the effect of enclosure filling material, among others.

We will also study a transmission line enclosure from an empirical perspective and explore anomalous impedance behavior in systems that can tell us a great deal about enclosure and driver imperfections. But first we must discuss the shift in resonant frequency that occurs when a driver is mounted on a baffle or in an enclosure.

3.1 DETERMINING f_{SB}

When suspended in free air, low-frequency acoustic radiation off the front and rear of a driver cone are out-of-phase and cancel each other. When placed on a large baffle or in a box, the front and back waves are isolated, substantially increasing low-frequency output. Air-mass loading on the cone also increases. This phenomenon, referred to as *mass reactance loading*, was discussed briefly in Chapter 2. The increase in effective cone mass lowers a driver's resonant frequency, raises its Q, and lowers its efficiency. Designing low-frequency enclosures using the T/S theory requires that we know driver resonant frequency, total Q, and equivalent volume. To use this theory optimally, however, we should correct the free-air parameters for the change in mass-reactance loading caused by a baffle or enclosure. Let's look at some examples.

3.1.1 INFINITE BAFFLE LOADING

In Chapter 2 I presented a formula for the mass reactance loading on a circular piston suspended in free air. The formula is repeated below:

$$M_{M1} = 0.566 S_D^{1.5} \qquad [3.1]$$

where S_D is in square meters. In the frequency region of driver resonance the wavelength in air is typically on the order of 10m, perhaps 50 times greater than the distance from the front to the back of the cone. At these low frequencies, air being displaced from the front rushes around the cone to fill the void at the rear, and thus, only one side of the cone is effectively mass loaded.

This cannot happen with an infinite baffle. Mass reactance loading doubles to account for the loading on both sides which are now isolated from each other. The total moving mass including air load under infinite baffle loading, denoted by M'_{MS}, is:

$$M'_{MS} = M_{MD} + 2M_{M1} = M_{MS} + M_{M1} \qquad [3.2]$$

Driver resonant frequency when mounted on an infinite baffle becomes

$$[3.3]$$

$$f_i = \frac{1}{\sqrt{M'_{MS} C_{MS}}} = \sqrt{\frac{M_{MD} + M_{M1}}{M_{MD} + 2M_{M1}}} \times f_{Sa}$$

For the 8″ driver used in the discussion of Section 2.8.1,

$$M_{M1} = 0.566 (0.0217)^{1.5} = 0.0018 \text{Kg} = 1.8\text{g}$$

Recalling that the total mass under free-air conditions is 22.3g, we get

$$M'_{MS} = 22.3 + 1.8 = 24.1\text{g}$$

where M'_{MS} is the total mass under infinite baffle loading, and

$$f_i = \sqrt{\frac{M_{MS}}{M'_{MS}}} \times f_{SA} = \left(\sqrt{\frac{22.3}{24.1}}\right) \times 23.4 = 0.96 \times 23.4 = 22.4\text{Hz}$$

where f_i is the driver resonant frequency under infinite baffle loading. Notice that the infinite baffle loading has lowered the resonant frequency by 4%.

Because the cone's moving mass is greater with infinite baffle loading, the motor will have more difficulty damping it, which implies that the electrical Q rises. Recall from Equation 2.13 that Q_{ES} is proportional to the square root of cone mass. The same is true for Q_{MS} (see Equation 2.5). Both Q_{ES} and Q_{MS}, and therefore Q_{TS}, will increase by the same amount that f_S decreases. The original value of Q_T was 0.4. Under infinite baffle loading it becomes:

$$Q_{TS} = 0.40/0.96 = 0.42$$

Finally, recall from Equation 2.37 that efficiency is inversely proportional to the *square* of the moving mass. The original efficiency of 0.56% for our 8″ driver will fall to

$$\eta_0 = (22.3/24.1)^2 \times 0.56 = 0.86 \times 0.56 = 0.48\%$$

This is a drop of 0.7dB. Using Equation 2.36, we can recalculate the power sensitivity:

$$S_P = 112.2 + 10 \times \log(.0048) = 112.2 - 23.2 = 89.0$$
$$\text{dB/1W/1m}$$

which is exactly 0.7dB lower than the result we obtained in Section 2.8.3.

3.1.2 BOX LOADING

The relations of the previous section can be applied to a driver mounted in a large wall or very large box. In this section we'll look at loading effects in more typical enclosures. Beranek[1] gives an approximate expression for acoustic mass loading on the *front* of a cone mounted in a medium-sized box (less than 8ft³). Converting this equation to mechanical mass we get the mass, reactance loading on the cone front, $M_{MR[front]}$:

$$M_{MR[front]} = 0.408\, S_D^{1.5} = 0.284\, D^3 \qquad [3.4]$$

where D is the effective cone diameter in meters. Determining the loading on the rear of the cone (the side of the cone facing the enclosure) is a bit more complicated. This loading depends on the ratio of the cone area to the baffle area.

Consider the following line of reasoning: If the piston has the same area as one side of the box, a minimal movement of the piston will pressurize the box. As the piston area becomes smaller, more movement of the piston is required to produce the same pressure. In practical loudspeakers a great deal of air must be accelerated by the cone to pressurize the box, placing an effective air-mass load on the rear of the cone. The equivalent mass loading, $M_{MR[rear]}$, is:

$$M_{MR[rear]} = 0.667 \times K_m \times S_D^{1.5} \qquad [3.5]$$

For rectangular baffles, K_m is given by:

$$K_m \cong 10^{-(0.462\beta + 0.057)} \qquad [3.6]$$

The symbol $\cong$ means that the equation is approximately correct, and β is the ratio of cone area to front baffle area. Let's apply the last three equations to a typical 12″ woofer. *Table 3.1* gives the free-air parameters for the considered woofer.

TABLE 3.1	
12″ WOOFER PARAMETERS	
f_{SA}	= 19.3Hz
Q_{MS}	= 1.71
Q_{ES}	= 0.39
Q_{TS}	= 0.32
M_{MS}	= 75.3g
V_{AS}	= 330 ltr
S_D	= 510cm² = 0.051m²
η	= 89.9dB/1W/1m

Based on its free-air parameters, a quasi-Butterworth QB3 vented box[2] alignment for this driver would require an internal volume of 139 ltr or about 4ft³ with the box tuned to 23.9Hz. Since this volume is less than 8ft³, we can use the above equations. From Equation 3.4 the air-mass load on the front of the cone is:

$$M_{MR[front]} = 0.408 \times (0.051)^{1.5} = 0.408 \times 0.0115 = 0.0047\text{Kg} = 4.7\text{g}$$

To determine the loading on the rear of the cone we must pick some enclosure dimensions.

Rectangular parallelepipeds with side dimensions in the ratio of 1.6:1.2:1 are said to be the most visually pleasing. This set of numbers is called the *golden ratio*. To determine the interior dimensions for a given volume using this ratio, call the shortest side the depth, d. Then:

$$(1.6d)(1.2d)(d) = V_B$$

or

$$1.92d^3 = V_B$$

and

$$d = \left(\frac{V_B}{1.92}\right)^{\frac{1}{3}}$$

The box height, h, and width, w, are then determined by multiplying d by 1.6 and 1.2, respectively. Exterior dimensions of the enclosure will differ slightly from the golden ratio depending on the thickness of the sidewalls. For large enclosure volumes the difference will be small. For our 139 ltr enclosure, 0.139m³, we have:

$$d = \left(\frac{0.139}{1.92}\right)^{\frac{1}{3}} = 0.42\text{m}$$

Height and width of the front panel dimensions are then 0.5m and 0.67m, respectively, for a frontal area of 0.335m². Then:

$$\beta = 0.051/0.335 = 0.152$$

$$K_m = 10^{-(0.462 \times 0.152 + 0.057)} = 10^{-0.127} = 0.746$$

$$M_{MR[rear]} = 0.667 \times 0.746 \times 0.0115 = 5.7\text{g}$$

In order to calculate M'_{MS}, first we must determine the mechanical mass, M_{MD}, by removing from M_{MS} the mass reactance load, M_{M1}, during the free-air tests. For the 12″ driver in *Table 3.1*, using Equation 3.1, we have:

$$M_{M1} = 0.566\,(0.051)^{1.5} = 6.5\text{g}$$

then

$$M_{MD} = M_{MS} - M_{M1} = 75.3 - 6.5 = 68.8\text{g}$$

and finally

$$M'_{MS} = M_{MD} + M_{MR[front]} + M_{MR[rear]}$$

$$M'_{MS} = 68.8 + 4.7 + 5.7 = 79.2\text{g}$$

Now we can compute the T/S parameters when the 12″ driver is mounted in a 139 ltr enclosure.

$$f_{SB} = \sqrt{\frac{75.3}{79.2}} \times 19.3 = 0.975 \times 19.3 = 18.8\text{Hz}$$

$$Q_{TB} = 0.32/0.975 = 0.33$$

Going back into the QB3 design charts[2] with the revised parameters, we get a box volume of 153 ltr and a tuning frequency of 22.6Hz. The box volume has increased by 10% and the tuning frequency is lowered by 5%. These are rather significant differences and point out the need to make these corrections as part of the design process. As we discussed above, the power sensitivity reduction is proportional to the square of the mass ratios. So our new sensitivity is:

$$S_p = 89.9 + 10 \times \log[(75.3/79.2)^2] = 89.9 + 10 \times \log(0.9) = 89.9 - 0.5 = 89.4 dB/1W/1m$$

The equations we have given so far in this section can be used to correct T/S designs in the *initial* design stages. However, using a test box, we can actually measure the mass reactance loading. In the next section we will see, among other things, how this is done.

3.2 DRIVERS IN CLOSED BOXES

In this section we will learn about low-frequency drivers in closed boxes using impedance data. A driver is placed in a closed box to isolate front and rear waves, thereby greatly increasing low-frequency output. The box volume is selected to obtain a desired Q and resonant frequency.

We already know from our discussions in Chapter 2 that the resonant frequency of a driver/box *system* is higher than that of the driver alone in free-air. This may seem contradictory based on the previous section, but the word "system" in the last sentence is important. The *effective* resonant frequency of the driver will fall slightly when placed on a large baffle or in a closed box as we have already seen, but the combination of the driver and the closed box will generally exhibit a higher resonant frequency than the driver alone. This is because the stiffness of the trapped air volume in the closed box adds to the suspension stiffness, increasing the total stiffness, thereby raising the resonant frequency of the "system."

Figure 3.1 shows impedance curves for an 8″ driver both in free air and mounted in a 64.5 ltr box. Although the impedance curve for the driver in a box peaks at a higher frequency, its general shape is similar to that of the free-air curve. In fact, if we didn't know the second curve was for a driver in a box, we might assume it to be the free-air curve for another driver with a higher resonant frequency. The driver/box combination simply looks like another driver with a slightly heavier cone and a stiffer suspension. Clearly, we can use this impedance data to calculate T/S parameters for the closed-box system using any of the procedures given in Chapter 2.

Figure 3.2 shows the low-frequency region of the closed-box impedance curve on an expanded scale. f_C denotes the closed-box system resonant frequency. The frequencies f_C, f_1, and f_2 are marked with an "o" on the plot. From this plot the needed impedance and frequency values are:

$$f_C = 39.9 Hz$$
$$R_{max} = 177.6\Omega$$
$$R_x = 36.7\Omega$$
$$f_1 = 32.8 Hz$$

$$f_2 = 48.4 Hz$$

From these values closed-box Qs can be calculated using the equations in Section 2.6 of Chapter 2. *Table 3.2* lists the T/S parameters derived from the free-air and closed-box impedance plots of *Figs. 3.1* and *3.2*.

Bracketed subscripts in *Table 3.2* with the trailing letter "c" refer to closed-box values. Notice that the closed-box Q, Q_{TC}, is very close to the critically damped value of 0.5. Its low-frequency response plot will look like that shown in Fig. 2.6c for Q = 0.5.

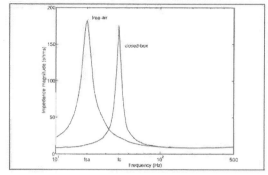

FIGURE 3.1: Free-air and closed-box impedance curves of 8″ driver.

We can learn a great deal more from the closed-box impedance plot. α is the ratio of driver equivalent volume to box volume; that is

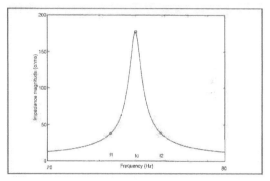

FIGURE 3.2: Expanded view of closed-box impedance curve.

$$\alpha = \frac{V_{AS}}{V_B} \qquad [3.7]$$

α is a critical parameter in the design of closed-box and vented systems. For example, we know from Chapter 1 of *LDC* that the ratio of driver free-air Q to closed-box Q is:

$$Q_{TC} = Q_{TS}\sqrt{\alpha + 1} \qquad [3.8]$$

α also predicts the upward shift in resonant frequency of the driver/box combination due solely to the increased stiffness (decreased compliance) of

TABLE 3.2
T/S PARAMETERS FOR THE 8″ DRIVER FROM *FIG. 3.2*

	Free-air	Closed-box system
$f_{S[C]}$(Hz)	19.3	39.9
R_E (Ω)	7.6	7.6
$Q_{MS[MC]}$	5.27	11.7
$Q_{ES[EC]}$	0.24	0.54
$Q_{TS[TC]}$	0.23	0.52
M_{MS}	21.3g	
V_{AS} (ltr)	222.4	
V_B (ltr)		64.5

the combined driver/box system.

$$f_C = f_{SA} \sqrt{\alpha + 1} \qquad [3.9]$$

Finally, α is a critical parameter in the T/S low-frequency direct-radiator alignment charts in Chapters 1 and 2 of *LDC*.

Substituting values from *Table 3.2* into Equations 3.8 and 3.9 we get

$$[3.10a]$$

$$Q_{TC} = 0.23 \times \sqrt{\frac{222.4}{64.5} + 1} = 0.23 \times \sqrt{4.45} = 0.49$$

$$f_C = 193 \times \sqrt{4.45} = 40.7 \text{Hz} \qquad [3.10b]$$

The predicted value of Q_{TC} is slightly lower and f_C is somewhat higher than the measured values. We have two problems here. The first is that we used the wrong values for Q_{TS} and f_S. The values in column 1 of *Table 3.2* are *free-air* values, which do not include the change in mass reactance loading encountered when going from free-air to closed-box loading. Due to the difficulty in measuring V_{AS} accurately, we may have a second problem with the calculated value of α.

Let's look at the first problem. We could obtain the correct driver frequency and Q values directly by first using the added mass technique to determine M'_{MS} and then using the equations from Chapter 2 to compute f_{SB}, the driver in-box resonant frequency, and the revised driver Qs, assuming the driver-mechanical compliance in the box has not changed from its free-air value. However, this approach is often difficult to implement, since attaching masses to the cone of a driver mounted in a box can be tricky. Removing the driver to mount the added mass and then remounting the driver will invariably lead to compliance shifts and cause additional error.

Fortunately, we can determine f_{SB}, the added mass reactance loading, and α indirectly with good accuracy by measuring T/S parameters in free air and in the box, and then performing a few simple calculations. The advantage of this approach is that we do not need to attach masses to the cone. Here is the procedure:

PROCEDURE 3.1

1. Measure the free-air values f_{SA} and Q_{ES} with any of the procedures given in Chapter 2.
2. Place the driver in an *unlined, well-sealed* box of known volume and measure the driver/box system parameters f_C and Q_{EC}.
3. Calculate the ratio of in-box to free-air masses as follows:

$$R_M = \frac{M'_{MS}}{M_{MS}} = \frac{f_{SA} \times Q_{EC}}{f_C \times Q_{ES}} \qquad [3.11]$$

4. Calculate α as follows:

$$\alpha = \frac{f_C \times Q_{EC}}{f_{SA} \times Q_{ES}} - 1 \qquad [3.12]$$

From Section 3.1 we know we only need the ratio of the masses to calculate all of the effective driver T/S parameters when mounted in an enclosure. Additionally, if we know the value of M_{MS}, we can compute the total effective moving mass, M'_{MS}, when mounted in the enclosure. Rewriting Equation 3.11 we get

$$M'_{MS} = R_M \times M_{MS} \qquad [3.13]$$

Substituting values from *Table 3.2* into Equation 3.11 we get

$$R_M = \frac{193 \times 0.54}{39.9 \times 0.24} = 1.088$$

The effective cone mass including the in-box mass reactance loading has increased by 8.8%. The new value of M'_{MS} is

$$M'_{MS} = 1.088 \times 21.3 = 23.2$$

which is an increase of 1.9g over the free-air value. The corrected in-box values for the T/S parameters of our driver are:

$$f_{SB} = \frac{f_{SA}}{\sqrt{R_M}} = \frac{19.3}{\sqrt{1.088}} = 18.5 \text{Hz}$$

$$Q_{MSB} = \sqrt{R_M} \times Q_{MS} = 1.043 \times 11.7 = 12.2$$

$$Q_{ESB} = \sqrt{R_M} \times Q_{ES} = 1.043 \times 0.24 = 0.25$$

In the above equations f_{SB} is the resonant frequency of our 8″ driver, including the effect of in-box mass-reactance loading, and Q_{MSB} and Q_{ESB} are the corresponding Q values. I emphasize again that f_{SB}, Q_{MSB}, and Q_{ESB} are the T/S parameters for the driver under box-acoustic loading. They are not the T/S parameters for the closed-box system. Having solved our first problem, the corrected T/S parameter driver values can now be used to predict the closed-box system T/S parameters using Equations 3.8 and 3.9.

$$f_C = \sqrt{\alpha + 1} \times f_{SB} = \sqrt{4.45} \times 18.5 = 39.1 \text{Hz}$$

$$Q_{MC} = \sqrt{4.45} \times Q_{MSB} = 2.11 \times 5.27 = 11.1$$

$$Q_{EC} = 2.11 \times Q_{ESB} = 2.11 \times 0.24 = 0.51$$

We still have a small discrepancy between the values calculated above and those listed in *Table 3.2*. This is undoubtedly due to an error in the compliance ratio, α. As discussed in Chapter 2, accurate measurement of V_{AS} is always problematic. Let's calculate α using Equation 3.12 and the values from *Table 3.2*.

$$\alpha = \frac{39.9 \times 0.54}{19.3 \times 0.24} - 1 = 3.65$$

This value for α differs by roughly 6% from the value of 3.45 calculated on the basis of the measured V_{AS}. Now,

$$\alpha + 1 = 4.65$$

and

$$\sqrt{\alpha + 1} = 2.15$$

With the revised value for α and the free-air T/S parameters corrected for air mass loading, the predicted closed-box parameters become:

$$f_C = 2.15 \times 18.5 = 39.9$$

$$Q_{MC} = 2.15 \times 5.27 = 11.3$$

and

$$Q_{EC} = 2.15 \times 0.24 = 0.52$$

These values are in excellent agreement with the measured in-box values given in column 2 of *Table 3.2*. Using data developed with Procedure 3.1, the measured value for the added mass loading on our 8″ driver example over its free-air value is 1.9g. The equations of Section 3.1.2 predict a value of 1.8g. This is in agreement and gives us confidence that the equations of Section 3.1 can be used to accurately predict in-box T/S driver parameters for design purposes.

This has been a long and somewhat involved section. Let's review what we have covered. First, we saw that the general shape of the impedance curve for a closed-box system was the same as that of a driver in free air. This led us to conclude that a closed-box system could be characterized by a set of T/S parameters like those used to characterize drivers in free air. We derived T/S parameters for the closed-box system using the same procedures developed in Chapter 2. Using Procedure 3.1, we learned how to correct driver free-air parameters to their "in-box" values. This required measuring two sets of T/S parameters, free-air and closed-box. These measurements also allowed us to calculate a very accurate value for the compliance ratio, α.

3.3 MEASURING THE EFFECT OF ENCLOSURE FILLING MATERIALS

Filling materials such as fiberglass, acoustic foam, and Dacron™, are placed in an enclosure to absorb internal reflections and suppress standing waves. The filling material also increases the effective acoustic volume of the enclosure and adds damping and mass loading to the cone, altering both driver- and system-resonant frequency, and Q. Let's see why this happens. The acoustic compliance of an enclosed volume is given by:

$$C_{AB} = \frac{V_B}{\rho \times c^2} \qquad [3.14]$$

where:
V_B = the enclosed volume in m³
c = speed of sound in the internal medium in m/s
ρ = the density of the internal medium in Kg/m³

The filling material changes both ρ and c, generally increasing the first and decreasing the second. The exact manner in which ρ and c change is complex, however, the ρc^2 product will always *decrease* over its unfilled value. This means that C_{AB} will *increase* over its unfilled value. Equivalently, you would need a larger *unfilled* box to give the filled-box value of C_{AB}. Thus, the filling material increases the acoustic volume of the box over its physical value.

An alternate form of Equation 3.14 gives us a physical interpretation of this effect.

$$C_{AB} = \frac{V_B}{\gamma \times P_0} \qquad [3.15]$$

where

P_0 = atmospheric pressure in N/m²
γ = a ratio of specific heats

What is γ? When an ideal gas is compressed, γ is the constant relating a small percentage change in pressure to a small percentage change in volume. That is:

$$\Delta P = -\gamma \Delta V \qquad [3.16]$$

where

ΔP = percentage change in pressure
ΔV = percentage change in volume

This equation should make physical sense because as we decrease the volume by pushing in on a driver cone (a negative change in V), the pressure inside the box pushing back on the cone increases (a positive change in pressure). Now the value of γ depends on the kind of gas and how the gas is compressed. If the gas is compressed rapidly, the temperature of the gas will rise rapidly, because the build-up of heat energy inside the box due to compression does not have time to dissipate into the surrounding air. This is called *adiabatic compression*, and the value of γ for air under adiabatic compression is 1.41. If the compression occurs very slowly, the heat build-up will dissipate into the surrounding atmosphere and the temperature of the gas will remain constant. This is called *isothermal compression*. The value of γ for air under isothermal compression is 1.

The compression (and rarefaction) of air in an acoustic wave is normally adiabatic because it occurs rapidly at audio frequencies. The filling material, however, adds thermal mass to the enclosed volume which tends to keep the temperature from changing during rapid compression or expansion. In other words, the filling material tends to change the compression from adiabatic to something closer to isothermal. γ also relates the acoustic volume of the enclosure, V_{AB}, to its physical volume, V_B.

$$V_{AB} = \frac{1.41 \times V_B}{\gamma_F} \qquad [3.17]$$

Under ideal conditions γ_F, the filled-box γ, could be as small as 1 for an increase in acoustic volume of 41% over that of an unfilled box. In practice, increases of 15%–20% are more typical. The following procedure, which is similar to Procedure 3.1, will allow us to evaluate the effect of filling material completely.

PROCEDURE 3.2:

1. Measure the free-air values f_{SA} and Q_{ES} with any of the procedures given in Chapter 2.
2. Place the driver in a *filled, well-sealed* box and measure the filled-box parameters f_{CF} and Q_{ECF}.
3. Calculate the ratio of filled-box to free-air cone masses:

$$R_{MF} = \frac{f_{SA} \times Q_{ECF}}{f_{CF} \times Q_{ES}} \qquad [3.18]$$

4. Calculate α_F, the filled-box α:

$$\alpha_F = \frac{f_{CF} \times Q_{ECF}}{f_{SA} \times Q_{ES}} - 1 \qquad [3.19]$$

5. Using the value of α obtained with Procedure 3.1 for the unfilled box, calculate the increase in acoustic volume:

$$\frac{V_{AB}}{V_B} = \frac{\alpha}{\alpha_F} \qquad [3.20]$$

An example will help make these equations clear. We will continue with our 8″ driver and 64.5 ltr test box example started in Section 3.2. The test box was filled with 50 oz of Dacron™ pillow stuffing. The filling material was fluffed and teased to maintain a uniform density. *Figure 3.3* shows the impedance curve obtained for the filled-box system. *Figure 3.4* clearly shows the dramatic impact the filling material has on system impedance. Both the unfilled- and filled-box curves are plotted together in this figure for comparison. Notice the impedance peak has dropped by more than a factor of three.

The critical values for determining T/S parameters for the filled-box system taken from *Fig. 3.3*

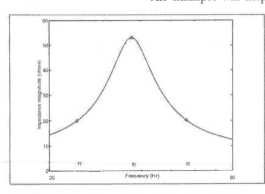

FIGURE 3.3: Impedance plot for 8″ driver in a filled box.

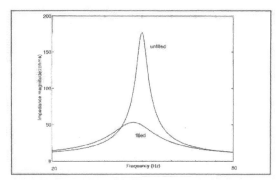

FIGURE 3.4: Comparison of unfilled and filled box impedance curves.

**TABLE 3.3
T/S PARAMETERS FROM *FIGS. 3.2 & 3.3***

	Free-air	Filled-box
$f_{S(CF)}$ (Hz)	19.3	37.3
R_E (Ω)	7.6	7.6
$Q_{MC[F]}$	5.27	3.11
$Q_{EC[F]}$	0.24	0.52
$Q_{TS(CF)}$	0.23	0.45
M_{MD} (g)	21.1	
V_B (ltr)		64.5

are listed below:

$$f_{CF} = 37.3 Hz$$
$$R_{max} = 53.05\Omega$$
$$r_0 = 53.05/7.6 = 6.98$$
$$\sqrt{r_0} = 2.64$$
$$R_x = 2.64 \times 7.6 = 20.1\Omega$$
$$f_1 = 24.7 Hz$$
$$f_2 = 56.3 Hz$$

Table 3.3 lists the T/S parameters calculated from the above data along with the free-air parameters obtained previously.

Using the values in *Table 3.3* and Equations 3.18–3.20 we get the following:

$$R_{MF} = \frac{f_{SA} \times Q_{ECF}}{f_{CF} \times Q_{ES}} = \frac{19.3 \times 0.52}{37.3 \times 0.24} = 1.12$$

$$\alpha_F = \frac{f_{CF} \times Q_{ECF}}{f_{SA} \times Q_{ES}} - 1 = \frac{37.3 \times 0.52}{19.3 \times 0.24} - 1 = 3.19$$

$$\frac{V_{AB}}{V_B} = \frac{\alpha}{\alpha_F} = \frac{3.65}{3.19} = 1.15$$

We see that the box filling has increased mass reactance loading from 8.8% to 12% of the unloaded cone mass. This happens because filling material immediately behind the cone tends to move with the cone. Also, the effective volume of the enclosure has increased by 15%. Some other interesting things have happened that will shed more light on what the filling material does. *Table 3.4* compares T/S parameters for the unfilled-box and filled-box conditions.

TABLE 3.4
CLOSED-BOX T/S PARAMETERS WITH
AND WITHOUT FILLING

	Unfilled box	Filled box
$f_{C[F]}$(Hz)	39.9	37.3
$Q_{MC[F]}$	11.7	3.11
$Q_{EC[F]}$	0.54	0.52

In all of the above tables and subsequent equations, the trailing "F" in the subscripts refers to filled-box values.

As you would expect, the system resonant frequency and electrical Q have slightly dropped. However, there is a very large drop in system mechanical Q. Let's see if we can figure out what is happening here. Our first step is to compute the effective filled-box *driver* parameters using the value of R_{MF} just determined. From the original free-air values from *Table 3.1* we get:

$$f_{SBF} = \frac{f_{SA}}{\sqrt{R_{MF}}} = \frac{19.3}{\sqrt{1.012}} = 18.2 \text{Hz}$$

$$Q_{MSB} = \sqrt{R_{MF}} \times Q_{MS} = 1.058 \times 11.7 = 12.4$$

$$Q_{ESB} = \sqrt{R_{MF}} \times Q_{ES} = 1.058 \times 0.24 = 0.254$$

Now we will use α_F to compute the filled-box T/S parameter values.

$$f_{CF} = \sqrt{\alpha_F + 1} \times f_{SBF} = \sqrt{3.19 + 1} \times 18.2 = 2.047 \times 18.2 = 37.3 \text{Hz}$$

$$Q_{ECF} = 2.047 \times 0.254 = 0.52$$

$$Q_{MSF} = 2.047 \times 5.27 = 10.8$$

Observe that the filled-box resonant frequency and electrical Q are perfectly consistent with the measured value of α_F. Once the driver parameters are corrected for box loading, the measured value of α_F accurately predicts system-resonant frequency, f_{CF}, and electrical Q, Q_{ECF}. But there seems to be a big problem with mechanical Q, Q_{MSF}. It does not match the filled-box system Q listed in *Table 3.4*. That's because we have not properly accounted for the damping introduced by the filling material.

Notice I used the symbol Q_{MSF} rather than Q_{MCF} for the computed mechanical Q of the filled-box system. Q_{MSF} is only that portion of the mechanical Q contributed to the system by the driver. We must also account for box losses introduced by the filling material. In order to do this we'll have to backtrack a bit.

Figure 3.5a shows an equivalent mechanical circuit model for a closed-box loudspeaker system. You are already familiar with the terms M_{MS}, C_{MS}, and r_{MS}. F_g, the driving force, is analogous to voltage, while velocity is the analog of current. C_{MB} is the mechanical compliance of the enclosed volume. r_{MB}, the mechanical equivalent of the acoustic absorption resistance, r_{AB}, will increase when the box is filled with absorptive material such as Dacron™ or fiberglass. r_{ML} represents losses due to enclosure leakage, which may be caused by poor sealing of the box joints or at the driver-mounting flange, or leakage through the driver surround or dust cap.

If the box is well-sealed with rigid walls and no internal filling material, r_{ML} becomes large and r_{MB} drops to zero. For this condition, *Fig. 3.5a* simplifies to the analogous circuit shown in *Fig. 3.5b*, where the series capacitors C_{MS} and C_{MB} were combined into a single capacitor, C_{MT}. Recall that series capacitors combine like parallel resistors and that the resulting capacitance (read compliance) is less than either of the original capacitors (compliances). This is consistent with the discussion in the opening paragraphs of this section in that a decreased compliance corresponds to an increased stiffness for the system. The combined compliance is computed as shown below.

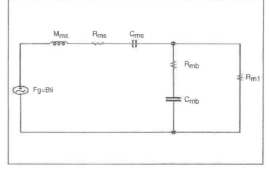

FIGURE 3.5a: Mechanical impedance analog for closed-box system.

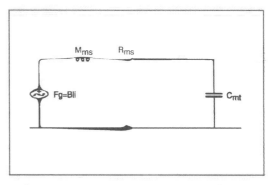

FIGURE 3.5b: Mechanical impedance analog for closed-box system with no box losses.

$$C_{MT} = \frac{C_{MS}C_{MB}}{C_{MS} + C_{MB}} \qquad [3.21]$$

Figure 3.5b corresponds to our first closed-box example, the box without filling. Because box losses are very low in this example, system mechanical Q is controlled largely by driver mechanical losses, and is accurately predicted by simply multiplying driver in-box Q, Q_{MSB}, by the term $\sqrt{\alpha+1}$. When box losses are significant, however, a different approach is needed.

Figure 3.5c shows the analogous mechanical circuit for the closed-box system with filling material, but no leakage. r_{MB} is now large and cannot be ignored. The two-series compliances could still be combined into a single compliance, and the resistors r_{MB} and r_{MS} could be added to form a single resistor, but it is instructive to keep them separate.

This allows us to assign a mechanical Q to the box. In particular, this Q will characterize box-absorption losses. (If leakage is significant a second box Q will be needed to account for the leakage losses.) The mechanical Q of the filled-box *system* is

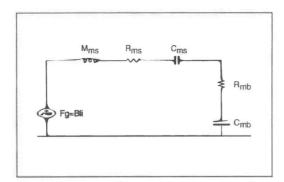

FIGURE 3.5c: Mechanical impedance analog for closed-box system with absorption losses.

now a function of both driver and box Qs. The two Q values combine just like resistors in parallel. That is:

$$\frac{1}{Q_{MCF}} = \frac{1}{Q_{MSF}} + \frac{1}{Q_{MB}}$$ [3.22]

where:

$$Q_{MB} = \frac{1}{2\pi f_{CB} C_{MB} r_{MB}}$$

Equation 3.22 shows us that the mechanical Q computed earlier represents just the driver contribution to mechanical losses. We can solve Equation 3.22 for Q_{MB}.

$$\frac{1}{Q_{MB}} = \frac{1}{Q_{MCF}} - \frac{1}{Q_{MSF}} = \frac{1}{3.11} - \frac{1}{12.4} = 0.24$$

Inverting this result:

$$Q_{MB} = 1/0.24 = 4.15$$

In this example, notice that the box Q due to absorption losses is much lower than the driver mechanical Q; therefore, the mechanical Q of the filled-box system is dominated by box losses. These losses caused the large drop in system impedance seen in *Figs. 3.3* and *3.4*. This strongly contrasts with the unfilled system where driver losses accounted for the entire system mechanical Q. Unfilled boxes with rigid walls have mechanical

Qs on the order of 50–100. From this last example it is clear that box-filling materials not only produce a significant increase in effective acoustic volume, but they also add to driver-cone mass-reactance loading and significantly increase acoustic absorption loss over that experienced with an unfilled box.

Summarizing the results of this section—measurement of resonant frequency and electrical Q under three conditions: free-air, unfilled closed-box, and filled closed-box—allows us to determine all the important T/S parameters for a closed-box system, including a full evaluation of the effect of the filling material. Amazingly, all of these data are derived from impedance measurements alone.

3.4 DRIVERS IN SINGLE-TUNED VENTED SYSTEMS

In this section we will learn about the impedance plot of simply vented systems. These systems, sometimes referred to as bass reflex systems, have a single tuned chamber that generally provides increased bass energy output relative to a closed-box system. The equations and results of this section also apply to bandpass systems with one sealed and one vented chamber. The vented and bandpass systems discussed in this section are shown in *Fig. 3.6*.

Figure 3.7a is a typical impedance plot for a vented system. The plot is characterized by two peaks with a valley between them. The height of the peaks, which are a function of system losses, are generally not the same. Referring to this plot, here are some new parameters we will need for discussion in this section:

f_B = box-tuning frequency
f_H = the frequency of the higher-frequency impedance peak
f_L = the frequency of the lower-frequency impedance peak
f_M = the frequency of the minimum impedance between the two peaks
h = the tuning ratio, f_B/f_{SB}
R_M = the impedance magnitude at f_M

In order to determine whether or not a given vented system has met its design goals, we need to measure the actual values of the box-tuning frequency, f_B; the compliance ratio, α; the tuning ratio, h; and the box Q. We also need to know the

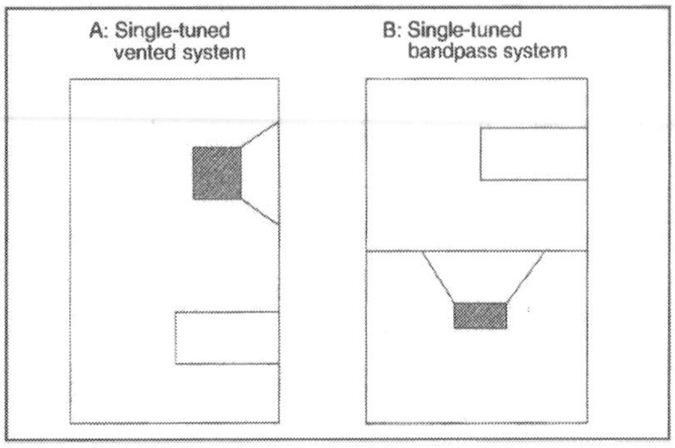

FIGURE 3.6: Vented (A) and bandpass (B) systems discussed in Section 3.4.

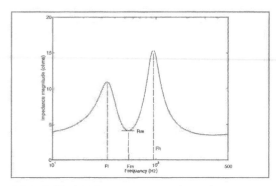

FIGURE 3.7a: MTM vented-box impedance plot.

in-box driver parameters, f_{SB}, Q_{MSB}, and Q_{ESB}. The following set of impedance measurements will allow us to calculate all of these parameters.

PROCEDURE 3.3: VENTED-SYSTEM IMPEDANCE ANALYSIS

1. From the vented-system impedance curve, record the values of f_H, f_L, f_M, and R_M.

2. Carefully seal the port and measure the closed-box resonant frequency, f_C.

3. Calculate the vented-system parameters using the formulas below:

$$f_B = \left(f_H^2 + f_L^2 - f_C^2\right)^{1/2} \quad [3.23]$$

$$\alpha = \frac{\left(f_H^2 - f_B^2\right)\left(f_B^2 - f_L^2\right)}{f_H^2 f_L^2} \quad [3.24]$$

4. Calculate the "in-box" driver T/S parameters as follows:

$$f_{SB} = \frac{f_H f_L}{f_B} \quad [3.25]$$

$$Q_{MSB} = \left(\frac{f_{SA}}{f_{SB}}\right)Q_{MS} \quad [3.26]$$

$$Q_{ESB} = \left(\frac{f_{SA}}{f_{SB}}\right)Q_{ES} \quad [3.27]$$

$$Q_{TSB} = \left(\frac{f_{SA}}{f_{SB}}\right)Q_{TS} \quad [3.28]$$

5. Calculate the box Q and tuning ratio as follows:

$$r_M = \frac{R_M}{R_E} \quad [3.29]$$

$$Q_B = \frac{h}{\alpha}\left(\frac{1}{Q_{ESB}(r_M - 1)} - \frac{1}{Q_{MSB}}\right) \quad [3.30]$$

$$h = \frac{f_B}{f_{SB}} \quad [3.31]$$

Notice that the parameter f_M does not appear in any of the above equations. If box losses are low (Q_B is large), voice-coil inductance small, and driver impedance measured directly at the voice-coil terminals without a crossover in place, f_M will generally be very close to f_B in value, and may be used as a first approximation to f_B. (If voice-coil inductance is large, f_M will fall below f_B.) Unfortunately, the impedance curve in the region around f_M is very shallow, so that it is often difficult to determine the exact minimum. Conversely, because the impedance peaks at f_L, f_H, and f_C are sharp, these frequen-

TABLE 3.5
DUAL 130mm DRIVERS: VENTED-SYSTEM AND FREE-AIR PARAMETERS

$f_{SA} = 64.2$Hz	$f_H = 94.2$Hz
$Q_{MS} = 1.82$	$f_M = 54.0$
$Q_{ES} = 0.39$	$f_L = 34.2$
$Q_{TS} = 0.32$	$f_C = 84.7$
$R_E = 6.32\Omega$	$R_M = 4.15\Omega$

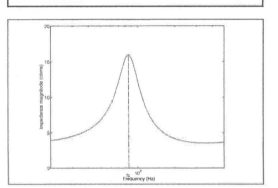

FIGURE 3.7b: MTM closed-box impedance plot.

cies can be measured accurately. The extra closed-box measurement of f_C permits a very accurate determination of f_B, α, and h.

The measured value for α includes the effect of any enclosure filling or lining material. You can use Procedure 3.3 to determine V_{AS}, but all filling or lining material must be removed to get an accurate result. If this is the case then:

$$V_{AS} = \alpha \times V_B$$

The total box Q for a simply vented system is given by

$$\frac{1}{Q_B} = \frac{1}{Q_L} + \frac{1}{Q_A} + \frac{1}{Q_P}$$

where Q_L and Q_A are leakage and absorption losses, and Q_P represents port losses. The above equations for f_{SB} and α assume that leakage losses are small. This means that the enclosure must be well-sealed, and leakage around the driver-mounting flange, surround, and dust cap must be small.

Where does f_M come into the picture? Equation 3.30 for the total box Q was derived under the assumption that f_M and f_B are approximately equal. If for some reason measuring f_C is not convenient, you can use f_M in place of f_B in all of the above equations. The accuracy of the assumption can be checked with the equation given below:

$$\frac{f_B}{f_M} = \sqrt{\frac{\alpha Q_B^2 - h^2}{\alpha Q_B^2 - 1}} \quad [3.32]$$

If this ratio is close to one, the assumption that f_M is close to f_B is good. If the ratio is off by a few percent, Equation 3.32 can be used to compute a corrected value for f_B, which can then be used to repeat the calculations.

You have just been bombarded with a mess of equations. I hope a couple of examples will make their use clear. *Figure 3.7a* is the measured impedance plot for a small monitor speaker using two 130mm mid-bass drivers in an MTM geometry. *Figure 3.7b* shows the impedance plot for the same system with the port vent sealed. Driver free-air parameters and data taken from the two impedance plots are listed in *Table 3.5*.

We begin by calculating f_B using Equation 3.23:

$$f_B = [94.2^2 + 34.2^2 - 84.7^2]^{1/2}$$
$$= [8873.64 + 1169.64 - 7174.09]^{1/2}$$
$$= [2869.19]^{1/2} = 53.6\text{Hz}$$

From Equation 3.24, the compliance ratio, α, is:

$$\alpha = \frac{\left(94.2^2 - 53.6^2\right)\left(53.6^2 - 34.2^2\right)}{\left(94.2^2\right)\left(34.2^2\right)}$$

$$\alpha = 0.99$$

The loaded driver parameters are now calculated using Equations 3.25–3.28.

$$f_{S_B} = \frac{f_H f_L}{f_B} = \frac{94.2 \times 34.2}{53.6} = 60.1\text{Hz}$$

$$Q_{MSB} = \left(\frac{64.2}{60.1}\right)(1.82) = 1.068 \times 1.82 = 1.94$$

$$Q_{ESB} = 1.068 \times 0.39 = 0.42$$

$$Q_{TSB} = 1.068 \times 0.32 = 0.34$$

Finally, calculate the tuning ratio, h, and box Q using Equations 3.29–3.31.

$$h = \frac{f_B}{f_{S_B}} = \frac{53.6}{60.1} = 0.89$$

$$r_M = \frac{R_M}{R_E} = \frac{4.16}{3.20} = 1.3$$

$$Q_B = \frac{0.89}{0.99}\left[\frac{1}{0.42(1.3-1)} - \frac{1}{1.94}\right] = 6.7$$

The calculated value of Q_B is quite close to the value of seven, which is typically assumed in trial designs detailed in alignment tables in Chapter 2 of *LDC*. The enclosure for this example was lined with "egg crate"-style acoustic foam. Notice that f_M is within one percent of f_B and could have been used in place of f_B in the above calculations. This system does not appear to correspond to any of the well-known alignments. For example, using the values of f_{SB}, Q_{TSB}, and Q_B, a QB3 alignment for this driver pair would require α and h values of 2 and 1.13, respectively.

TABLE 3.6 180mm DRIVER: VENTED-SYSTEM AND FREE-AIR PARAMETERS

$f_{SA} = 25.7\text{Hz}$	$f_H = 67.6\text{Hz}$
$Q_{MS} = 1.81$	$f_M = 43.2$
$Q_{ES} = 0.29$	$f_L = 16.6$
$Q_{TS} = 0.25$	$f_C = 53.0$
$R_E = 5.27\Omega$	$R_M = 5.77\Omega$

Our second example involves a 180mm driver mounted in a 15 ltr vented box. There is no lining or filling material in the box, so we should expect box Q to be higher than in the first example. Vented- and closed-box impedance plots for this example are shown in *Figs. 3.8a* and *3.8b*. Driver free-air parameters and measured vented-box parameters are given in *Table 3.6*.

First, calculate f_B using Equation 3.23:

$$f_B = [67.6^2 + 16.6^2 - 53.0^2]^{1/2} = [2036.3]^{1/2} = 45.1\text{Hz}$$

From Equation 3.24, the compliance ratio, α, is:

$$\alpha = \frac{\left(67.6^2 - 45.1^2\right)\left(45.1^2 - 16.6^2\right)}{\left(67.6^2\right)\left(16.6^2\right)}$$

$$\alpha = 3.54$$

Now, calculate the loaded driver parameters using Equations 3.26–3.28.

$$f_{SB} = \frac{f_H f_L}{f_B} = \frac{67.6 \times 16.6}{45.1} = 24.9\text{Hz}$$

$$Q_{MSB} = \left(\frac{25.7}{24.9}\right)(1.81) = 1.032 \times 1.81 = 1.87$$

$$Q_{ESB} = 1.032 \times 0.29 = 0.30$$
$$Q_{TSB} = 1.032 \times 0.25 = 0.26$$

Finally, calculate the tuning ratio, h, and box Q using Equations 3.29–3.31.

$$h = \frac{f_B}{f_{S_B}} = \frac{45.1}{24.9} = 1.81$$

$$r_M = \frac{R_M}{R_E} = \frac{5.77}{5.27} = 1.095$$

$$Q_B = \frac{1.81}{3.54}\left[\frac{1}{0.3(1.095-1)} - \frac{1}{1.85}\right] = 17.7$$

There are several points of interest in this example. First, as we expected, the box Q is quite high. Our first clue to this high Q lies in the small difference between R_E and R_M. Second, f_M and f_B differ by 4.4%, with f_M being lower. As discussed earlier, this is caused by a large voice-coil inductance. Using f_M in place of f_B in this example will lead to large errors.

As with our first example, the measured alignment does not correspond to any of the common alignments in *LDC*. We were trying for a QB3 alignment with this driver that requires a tuning ratio of 1.5 and an α of 4.1, assuming a Q_B of 7. In sizing the enclosure for this example, V_{AS} was taken from manufacturer's data. The V_{AS} for this sample is smaller, resulting in a lower value for α. Also, the port tube was inadvertently cut too short. The data quickly revealed the errors in this realization.

Comparing the two examples brings out another point. The mass-reactance loaded driver resonant frequency, f_{SB}, in the first example, dropped by 6.4%, relative to f_{SA}. Remember from our discussion in Section 3.1 that f_{SA} drops by the square root of

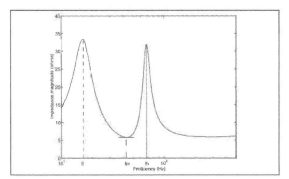

FIGURE 3.8a: 180mm driver vented-box impedance plot.

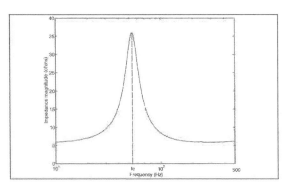

FIGURE 3.8b: 180mm driver closed-box impedance plot.

the mass ratio. This means that the radiation mass loading on the cones of the MTM driver pair increased by about 13%! In the second example, f_{SA} dropped by only 3.2%, corresponding to a mass increase of 6.5%.

Why is the loading so much greater in the first example? There are two reasons. First, mutual coupling between two closely-spaced drivers at low frequencies causes the drivers to load one another. The local sound pressure level produced by one driver actually pushes on the second driver, and vice versa. A second effect is related to the enclosure depth. The MTM enclosure is only 5″ deep. The proximity of the rear wall to the drivers increases the backside loading on the drivers over that of a much deeper enclosure. Enclosure depth is not taken into account in the approximate loading equations given in Section 3.1, but it can be significant.

3.5 AN EMPIRICAL LOOK AT TRANSMISSION-LINE IMPEDANCE

Unlike closed-box and vented-box systems, the transmission line (TL) lacks a widely accepted formal mathematical model. Yet this method of low-frequency driver loading has a devoted (even if small) following, especially in the DIY community.

There is an excellent summary of transmission-line design approaches in Chapter 4 of *LDC*. They are largely empirical and based on successful past designs. In Chapter 7 we will determine the speed of sound in a stuffed line using time-domain techniques. Here we will examine one transmission line impedance example.

Figure 3.9a is an impedance plot for an unstuffed transmission line. The system is com-

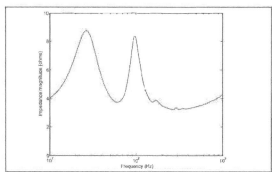

FIGURE 3.9a: Transmission line impedance with stuffing removed.

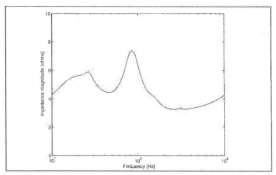

FIGURE 3.9b: Stuffed transmission line impedance.

prised of a 6.5″ mid-bass driver placed at one end of a 40″ line of constant cross-sectional area equal to $1.25S_D$. At first glance this looks just like the curve for a single-tuned vented system. (In fact, one noted speaker designer insists that TL performance is no better than a properly constructed vented system.[3]) The small peak at 170Hz is the result of a parallel resonance formed between the compliance of the small volume immediately behind the driver and the inductance of the line. The small wiggle at 300Hz is caused by a side-panel mechanical resonance.

After taking the data for *Fig. 3.9a*, the line was filled uniformly with 20 oz of Acousta-Stuff™, a nylon polyamide sound-damping fiber. The new impedance plot is shown in *Fig. 3.9b*. The upper impedance peak at 99Hz has been damped slightly, but the lower impedance peak at 27Hz has been reduced by 50%. Similar to our earlier closed-box analysis, this implies a substantially increased load on the driver at 27Hz, providing reduced cone excursion over a comparable vented box.

3.6 ANOMALOUS IMPEDANCE DATA

In the examples given in Sections 3.1–3.4, the impedance data is well-behaved. The curves look very much like the T/S models would predict; there are

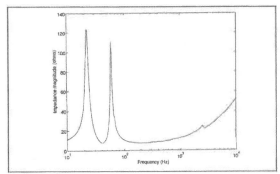

FIGURE 3.10a: Anomalous impedance Example No. 1.

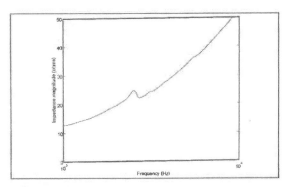

FIGURE 3.10b: Expanded view of *Fig. 3.10a.*

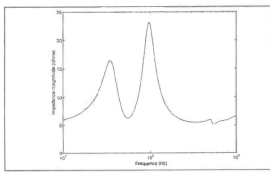

FIGURE 3.11: Anomalous impedance Example No. 2.

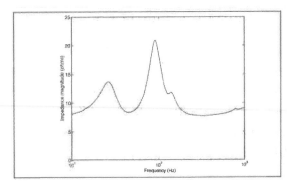

FIGURE 3.12: Anomalous impedance Example No. 3.

no strange wiggles or glitches in the plots. From earlier discussions, however, we know that all mechanical and acoustical loading on a driver is reflected back into its impedance data. Acoustical or mechanical energy coupled from a driver into the enclosure walls or the enclosed volume can produce cabinet-wall vibrations, internal reflections, and internal standing waves. We should expect that these undesirable effects would be reflected in the impedance data and that is the case.

We have already seen a hint of this in the TL example in Section 3.5. In this section we will examine four examples of anomalous impedance. Each case requires a bit of detective work to determine what is causing the impedance anomaly. The examples are not exhaustive, but they should give you a feel for how to proceed with your own investigations when you encounter anomalous impedance.

3.6.1 EXAMPLE NO. 1

Figure 3.10a is an impedance plot for a 12″ stamped frame woofer in a vented box. We are concerned with the "glitch" in the curve between 2–3kHz. *Figure 3.10b* is an expanded view of the glitch which is centered around 2.7kHz. Standing waves at 2.7kHz are highly unlikely, since the cabinet walls are lined with 2″ acoustic foam which is almost 100% absorbent at that high frequency.

Your first thought should be to look for some reflecting surface. A wavelength at 2.7kHz is about

5″ long. The webs of this driver's basket are rather wide and rest about 2.5″ behind the cone over its central area. I suspected that reflections off the basket were returning to the cone. They would return in phase at 2.7kHz. Lining the basket webs with 1/2″-thick foam completely eliminated the glitch. Reflections off the driver basket and cabinet walls can pass through the cone and interact with the primary radiation off the cone front. This can produce comb-filter response—much like what is discussed in Chapter 4—in connection with floor reflections.

3.6.2 EXAMPLE NO. 2
Impedance data for an MTM arranged pair of 6.5″ mid-bass drivers in a 30 ltr vented enclosure is plotted in *Fig. 3.11*. A glitch can be seen in this plot centered around 500Hz. This could be caused by a standing wave or an enclosure panel resonance. Further thinking made unlikely the standing-wave postulate for two reasons. First, no single dimension of the enclosure corresponded to a half-wavelength at 500Hz. (See Chapter 4 for a complete discussion of standing-wave modes in rectangular enclosures.) Secondly, it is actually somewhat difficult to excite standing waves in an enclosure with two separated in-phase sources rather than with a single source. There is an optimum location in any enclosure for exciting each standing wave mode. Both drivers cannot be in the optimum location and thus they will tend to fight each other.

With this reasoning in mind, the enclosure walls were scanned with a light finger touch while driving the system to a fairly high SPL. This scan revealed that both side panels were vibrating at 500Hz.

3.6.3 EXAMPLE NO. 3
Figure 3.12 shows impedance data for a home-theater loudspeaker. It is a tower design with four 5.25″ metal-cone mid-bass drivers in a vented MMTMM configuration. In addition to the double-peaked curve expected from a vented system, there is a third, smaller peak at 138Hz.

The tower is almost 6′ tall, but it has a false bottom. The active internal height is roughly 4′ 6″. This height will support a 125Hz standing wave, but this is not the culprit. The internal width and depth of the tower are 8″ and 10″, respectively. I placed an 8″ × 10″ piece of 1″ MDF just below the tweeter for internal stiffening.

Then I drilled a 5″ diameter hole in the center of this piece to connect the upper and lower volumes of the enclosure.

This stiffener effectively created a double-tuned system with two chambers and two resonant frequencies. The hole in the stiffener acted like an inductance connecting the two chambers. Removing the stiffening piece eliminated the 138Hz peak. A new stiffening piece with a larger rectangular cutout cured the problem.

3.6.4 EXAMPLE NO. 4
This example is an even more flagrant case of improper internal bracing. The system in question used a 12″ cast-frame woofer in a vented enclosure.

The impedance plot is shown in *Fig. 3.13*. Again we see the expected double-peaked curve plus a third peak at 120Hz. The cabinet for this system was rather deep. The manufacturer added a side-to-side panel brace that ran three-quarters of the internal cabinet height, dividing the enclosure into two almost equal halves.

I cut one 9″ diameter hole in the brace just behind the driver, resulting in a double-chambered system like that of Example No. 3. In this case, the resonance at 120Hz caused a sharp response dip of

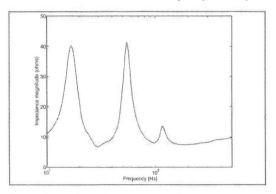

FIGURE 3.13: Anomalous impedance Example No. 4.

almost 3dB at this frequency. I cut several large rectangular holes in the interior bracing panel to eliminate the double-chamber effect.

3.6.5 DISCUSSION

All of the examples in this section come from measurements on real systems. *None of the data is simulated.* Uncovering impedance anomalies requires very high resolution impedance data. This data is most easily obtained with PC-based data acquisition systems discussed in Chapter 7.

It is possible, however, to find these anomalies with a careful manual scan using a sine-wave oscillator and the voltage-divider technique for measuring impedance. Scanning slowly and uniformly over a suspicious frequency range while watching the AC voltmeter scale, you will notice that most of the time, the meter needle will also move slowly and uniformly—either up or down—depending on the shape of the impedance curve. When an anomaly is reached, however, the needle will jump or move rapidly. The total needle motion will be small, but the suddenness of it will be readily apparent.

You can bracket the frequency extent of the anomaly with a steady hand on the oscillator tuning knob. You do not need a detailed plot of the impedance data through the anomaly. The most important data is the frequency range. You then must determine what caused the anomaly and decide whether or not it is worth rectifying.

REFERENCES
1. Beranek, L., *Acoustics*, Acoustical Society of America, 1996, Chapter 8.
2. Dickason, Vance, *The Loudspeaker Design Cookbook*, Audio Amateur Press, 1991, Chapter 2, Table 2.5.
3. Dickason, Vance, *The Loudspeaker Design Cookbook*, Audio Amateur Press, 1991, Chapter 4, p.

73. (Dickason quotes Martin Colloms here with no reference cited.)

CHAPTER 4

ACOUSTICAL TESTING OF SINGLE DRIVERS

4.0 INTRODUCTION

In this chapter we will thoroughly examine the factors influencing the accuracy of acoustical tests of individual drivers. The primary emphasis will be on obtaining accurate frequency-response data because it can be obtained with relatively simple instrumentation. More sophisticated tests for phase response, transient response, acoustic-phase center, and distortion, for example, will be covered in Chapter 7, which deals with PC-based acoustical and electrical testing systems. Along with impedance data, frequency-response data on individual drivers is generally used to evaluate drivers for a particular application and to develop data for crossover design.

After describing the frequency-response curve, we will discuss microphone types and explain which are best for loudspeaker testing. We will also examine the effect of the acoustic environment and baffle geometry on measurements and detail how to use various test techniques and test signals to mitigate environmental effects. Many examples will be provided to illustrate the material in this chapter.

4.1 THE FREQUENCY-RESPONSE PLOT

Figure 4.1 shows the setup for a frequency-response test in its simplest form. A sine-wave oscillator covering the audio-frequency range (20Hz–20kHz) drives a power amplifier, which in turn powers the driver under test (DUT). The DUT is placed on a baffle appropriate to the test. (I will discuss the effect of baffle geometry on driver response shortly.) The output of a microphone placed in front of the DUT is amplified and fed to a wideband AC voltmeter with scales calibrated in volts and decibels. The preamplifier may have an optional high-pass filter to reject low-frequency background noise generated by HVAC equipment.

Multiple frequency points are selected with the oscillator, the meter readings are recorded, and the results are plotted on a graph. Ideally, the frequency response comprises only the direct-field response of the driver or loudspeaker system, including the effect of baffle geometry. Ambient noise and late arrivals of acoustic energy from reflecting surfaces can corrupt the measurement. Much of the art of loudspeaker measurement deals with recognizing these errors and minimizing or eliminating them from the frequency-response data.

Several important features of the test and the resulting plot should be noted. First, the voltage amplitude applied to the driver voice coil is kept *constant* as frequency is varied. Second, the microphone senses the *sound-pressure level (SPL)* produced by the driver and converts it into a voltage for amplification and readout. Thus, the fre-

quency-response curve is a plot of SPL versus frequency at a constant-drive voltage. For dynamic drivers, the principal frequency-response plot—the on-axis response—is measured with the microphone placed along the extended voice-coil axis, typically at a distance of 1m (40″) from the baffle.

Using this axis as a reference, off-axis response data in both the horizontal and vertical planes is taken to build a picture of the driver's polar response (*Fig. 4.2*). Polar response is an important consideration in loudspeaker system design when selecting the crossover frequency between low- and high-frequency drivers.

Wideband, high-impedance AC voltmeters are particularly convenient instruments for interpreting the microphone output. These meters typically have switched full-scale sensitivities ranging from 1mV–300V. Bandwidths run from a few hertz to 1–4MHz. Both average-responding rectifier and true RMS meters are available, with the latter being generally more expensive. The Hewlett-Packard 403B and HP3100 are examples of the former and the latter, respectively.

These meters are often available on the surplus market at reasonable prices and are highly recommended. They have both voltage scales and decibel scales. The decibel scales are especially useful for loudspeaker testing. Many digital multimeters (DMMs) have AC scales, but they usually have very limited frequency response, often going as high as only a few hundred hertz. Before you use a DMM for driver testing, check its specifications to make sure that, as a minimum, it covers the entire audio band with better than 1dB flatness.

Before going any further I should define SPL. Sound pressure is the incremental variation in ambient atmospheric pressure caused by an acoustic wave. In the hearing process, the ear responds to this incremental pressure. Pressure is defined as force per unit area. The unit of pressure in the SI system is called the pascal. One pascal is equal to a pressure of one newton per square meter. Microphone sensitivity is typically quoted in units of millivolts per pascal (mV/Pa). Thus, a microphone with a sensitivity of 10mV/Pa will output 10mV in the presence of an incremental pressure of one pascal.

Because the human ear responds to sound pressure logarithmically, SPL is defined on a decibel scale. When using such a scale, you must choose a

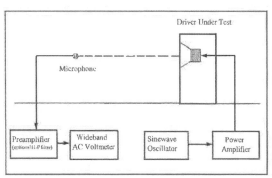

FIGURE 4.1: Setup for loudspeaker frequency response measurement.

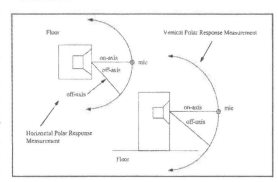

FIGURE 4.2: Microphone positions for polar response measurements.

0dB reference level. The threshold of audibility at 2kHz for the average young adult is 20µPa. This value is commonly used as the reference pressure level:

$$P_{REF} = 20\mu Pa = 20 \times 10^{-6} Pa$$

Then:

$$SPL = 20\log_{10}\left(\frac{P}{P_{REF}}\right) = 20\log_{10}(P) - 20\log_{10}(20 \times 10^{-6}) dB$$

$$SPL = 20\log_{10}(P) + 94dB \qquad [4.1]$$

where:

$$P = \text{sound pressure in pascals}$$

In Equation 4.1 the common base 10 log is used. From this equation notice that a pressure of one pascal corresponds to an SPL of 94dB since $\log_{10}(1) = 0$. For comparison, normal atmospheric pressure is about 10^5Pa, or 194dB above P_{REF}.

The frequency-response curve is normally plotted on linear decibel magnitude and logarithmic frequency scales. The plot can be in terms of absolute or relative SPL. In the case of relative frequency-response plots, a reference frequency must be chosen (1kHz is a common choice). The response at this frequency is arbitrarily set to 0dB. The plot will then show the departure of the frequency response from the reference point at other frequencies.

A typical frequency-response plot is shown in *Fig. 4.3* in both absolute and relative form. Notice that the shape of the curve is the same in both plots; only the decibel scale has changed. One advantage of the relative plot is that the microphone sensitivity may be unknown as long as the microphone response is known to be reasonably flat. The reading at the chosen reference frequency is recorded and then subtracted from all subsequent readings to get the relative response.

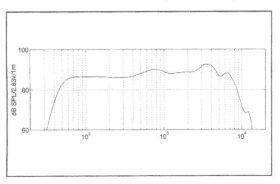

FIGURE 4.3a: Mid-bass driver, absolute frequency response.

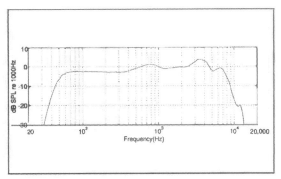

FIGURE 4.3b: Mid-bass driver, relative frequency response.

4.2 MICROPHONES FOR LOUDSPEAKER TESTING

A good microphone is at the heart of any acoustic measurement system. What constitutes a good microphone for loudspeaker testing? In my opinion, there are four key requirements:

1. Flat free-field frequency response
2. High, undistorted SPL capability
3. Good measurement repeatability
4. Long-term stability

The first two requirements are important for relative frequency-response measurements, and the last two are needed for absolute SPL measurements.

To meet frequency-response accuracy requirements for modern loudspeaker system design, a test microphone should have a *free-field frequency response* of at least 20Hz–20kHz. Ideally, the response between 20Hz–10kHz should be within ±1dB. Above 10kHz this requirement can be relaxed to ±2dB. Of course, the flatter the better. "Free-field" means the microphone's presence should not change the pressure it is measuring. That is, the microphone should measure the SPL that would exist at the point of measurement if the microphone were not there. I'll say more about the free-field later.

Assessing loudspeaker power response and distortion requires a test microphone that can handle SPLs of 120–130dB without distortion.

Repeatability refers to the variation of microphone characteristics from turn-on to turn-on. If you make a set of measurements one day, you'll probably need to repeat those measurements the following day. Of course, atmospheric conditions (temperature and humidity) and control of test geometry can also affect repeatability, but the microphone should not. Ideally, short-term repeatability should be ±0.3dB or better.

Long-term stability refers to a microphone's aging traits. There may be long periods when the microphone is not in use, sitting on a shelf under varying temperature and humidity conditions. The characteristics of a good microphone will not change after long shelf periods. Long-term changes in microphone sensitivity should not exceed ±0.5dB.

One microphone characteristic not on my list is *self-noise*. Top-quality recording and laboratory test microphones have self-noise levels on the order of 15dBA. (Noise levels in a "quiet" room can be on the order of 35–40dBA.) Since loudspeaker testing is typically conducted at reference SPLs of 90dB, higher noise levels can be tolerated. Sacrificing the noise specification can reduce microphone cost.

At the time of this writing, laboratory-grade microphones typically cost $3,500–4,000 US, although one brand sells for about $1,600 US. Instrument-grade electret-capacitor microphones are priced in the range of $500–$1,500 US. These prices are somewhat misleading, however, since the above microphones usually require a preamp to bring the microphone output voltage up to a readable level. Good preamps can add several hundred dollars to the cost. Some relatively inexpensive electret-condenser mikes are available with acceptable frequency response and self-noise levels of 25–35dBA. One such microphone, Mitey Mike II, is available from Old Colony Sound Laboratory (OSCL).[1]

4.2.1 MICROPHONE CLASSIFICATION

Microphones convert acoustic energy into electrical energy. They can be classified in terms of the particular property of the acoustic pressure field to which they respond. They can also be classified in terms of their directional properties and their transduction principles.

In the first category, microphones fall into three classes; pressure, pressure-gradient or velocity, and combinations of the two. In the second category, we have omnidirectional, bidirectional, and unidirectional. The last category comprises carbon, crystal or ceramic, dynamic, and capacitor. Carbon and crystal microphones have very ragged and limited frequency responses and are not suited for loudspeaker testing; therefore, they are not appropriate for further discussion.

A *pressure* microphone responds to changes in sound pressure. Only one side of the diaphragm is exposed to the incident acoustic wave and the back side is terminated in a closed cavity, much like a closed-box loudspeaker. A pressure microphone is inherently *omnidirectional*. Both capacitor and dynamic omnidirectional mikes are available.

In a *pressure-gradient* microphone, both sides of the diaphragm are exposed to the acoustic wave. In this case, diaphragm motion is caused by a pressure *difference* across the diaphragm. Pressure-gradient mikes are also referred to as velocity microphones, since particle velocity is proportional to the pressure gradient. These mikes are inherently *directional*.

A plane acoustic wave normally incident to the diaphragm surface will produce maximum differential pressure, and thus maximum electrical response. On the other hand, a plane wave propagating tangentially to the diaphragm will exert zero differential pressure and thus zero output voltage.

In all directional mikes the rear cavity is open to the atmosphere. The polar response is controlled by the way in which the incident wave is routed to the backside of the diaphragm. The length of the rear cavity, the number of openings along that length, and possibly some acoustical filtering elements determine the directivity pattern of the mike. Pressure-gradient microphones are made using either dynamic or capacitor transduction.

4.2.1.1 DYNAMIC MICROPHONES

Dynamic microphones are also referred to as moving-coil microphones. They can be viewed as miniature loudspeakers because like a loudspeaker, they have a diaphragm, a voice coil, and a magnetic circuit. Sound waves striking the diaphragm move the coil in the magnetic field to generate a voltage proportional to the incident sound pressure. Recall from the discussion in Chapter 2 that the voltage (EMF) generated across a coil moving in a magnetic field is proportional to the coil velocity. For a sinusoidal plane wave, pressure and velocity are $90°$ out-of-phase so that a dynamic microphone has an inherent phase shift of $90°$. However, the voltage magnitude is still proportional to the incident sound pressure.

Since the output voltage is proportional to diaphragm velocity, a dynamic microphone will not respond to a DC or static pressure increment. Like a closed-box loudspeaker, response will fall off at 12dB/octave below resonance. Proper resonance damping coupled with rear cavity tuning can often somewhat extend response, but low-frequency response will generally be inferior to that of capacitor microphones.

A dynamic microphone's diaphragm/voice-coil assembly mass limits high-frequency response, which is also usually inferior to that of capacitor microphones. Furthermore, changes in response with changes in temperature, atmospheric pressure, and humidity tend to be much higher relative to capacitor microphones. For example, sensitivity changes with temperature are typically on the order of 0.05dB/°C in good-quality dynamic mikes.

4.2.1.2 CAPACITOR MICROPHONES

In its simplest form, a capacitor microphone consists of a very thin diaphragm insulated from and placed a small distance in front of an electrically conducting backplate. The diaphragm and backplate form a capacitor. A DC polarizing potential on the order of a few hundred volts is applied to the backplate through a large series resistance to charge the capacitor. The large resistance assures that the capacitor will operate in a constant-charge mode. Under this condition, varying sound pressure deflects the diaphragm, changing the capacitance and producing an electrical signal directly proportional to the incident SPL.

Unlike dynamic microphones, the electrical output of a capacitor microphone is in-phase with the applied pressure. (Actually, laboratory capacitor mikes usually invert polarity, but this is easily corrected in the test setup.) The electret microphone is another form of capacitor microphone in which the polarizing high-voltage supply is replaced by an electret that holds a permanent electric charge.

In theory, capacitor microphones will respond to a static DC pressure increment. In practice, AC coupling is often used in the preamp to limit low-frequency response, but flat response down to 5Hz is common in lab-grade microphones. With no need for a voice coil, the diaphragm of a capacitor microphone can be made extremely light, allowing high-frequency extension to 20kHz and beyond.

Response of some 0.25″ diameter lab-grade capacitor microphones extends out to 100kHz. Capacitor microphones generally display much lower sensitivity to environmental effects than dynamic microphones. For example, a studio-grade capacitor microphone may have a temperature sensitivity of ±0.02dB/°C. For lab-grade microphones the number is ±0.007dB/°C.

4.2.2 WHICH MICROPHONE SHOULD YOU USE?

The answer may be obvious from the preceding discussion, but directly stated, pressure-sensitive capacitor microphones are preferred for loudspeaker testing. Relative to capacitor microphones, dynamic microphones exhibit higher environmental sensitivity, lower bandwidth, poorer transient response due to a heavier diaphragm, and a $90°$ phase shift. Compared to omnidirectional microphones, directional microphones have rougher frequency response and a low-frequency proximity effect. A directional microphone's response depends on the sound source's size and its distance from the source. As a source moves closer to a directional microphone, low-frequency response increases.

Later in this chapter we will describe the use of near-field measurements for loudspeaker testing where the microphone is placed within 0.5" of a driver's cone. For this test, a directional microphone will give an erroneous exaggerated bass response.

Omnidirectional microphones are capable of very flat response over the entire audio spectrum because only the front of the diaphragm is exposed to the incident wave, eliminating the phase cancellations experienced with directional microphones. However, there is one caveat. Omnidirectional microphone response to a plane wave becomes increasingly directional as the diameter of the diaphragm becomes comparable to the wavelength of the frequency being measured. Low frequencies "flow" past the diaphragm without incident. High frequencies, however, cannot bend around the microphone structure. They pile up on the diaphragm surface causing a rise in pressure that is higher than the pressure in the surrounding field. (This is one manifestation of acoustic diffraction.)

A pure-pressure microphone will show a rising response to a plane wave starting at a frequency whose wavelength is roughly ten times the diameter of its diaphragm. It is possible, however, to control diaphragm resonance and damping to counteract this effect and produce a flat response. This is the characteristic of a *free-field* pressure microphone. This type of microphone measures the pressure that would exist at the sampling point if the microphone were not there. Omnidirectional microphones for recording and broadcast applications are generally calibrated for free-field operation.

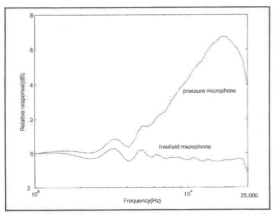

FIGURE 4.4: Free-field responses of pressure and free-field microphones.

Figure 4.4 shows the response to a plane wave of two ½" lab-grade capacitor microphones, one pure-pressure type and one free-field type. From the above discussion, the frequency at which the rising response should begin is:

$$f = \frac{c}{10D} = \frac{1130}{10 \times \frac{0.5}{12}} = 2712\,Hz$$

This is just what we see in the plot. Many engineering and laboratory test applications require a pressure microphone, but for loudspeaker testing a free-field microphone should be used. See Appendix A for a list of currently available microphones for loudspeaker testing.

4.3 THE ACOUSTIC ENVIRONMENT AND ITS EFFECT ON FREQUENCY RESPONSE

Unlike most electrical measurements, acoustic measurements are strongly influenced by their environment. It is a relatively straightforward matter to measure a battery voltage to three figures past the decimal place with today's DMMs. Although the meter is slightly influenced by temperature and lead resistance, measurements to 0.1% accuracy are commonplace. By contrast, an accuracy of $\pm1dB$ ($\pm10\%$) in acoustic measurements is considered excellent, with a more common accuracy of ±2 or $\pm3dB$.

In this section, I will discuss several aspects of the acoustic environment that influence the accuracy of frequency-response measurements. I will also explain the effect of baffle geometry on frequency response. At first, the following discussion may seem somewhat disjointed, but it will all come together in Section 4.4.

4.3.1 CLASSIFYING ACOUSTIC ENVIRONMENTS

There are several ways to classify an acoustic environment. For example, the environment can be anechoic (literally, without echo), semireverberant, or reverberant. Most of you are somewhat familiar with the use of *anechoic chambers* for acoustical testing. An anechoic chamber is an acoustically dead room with sound-absorbing wedges on all six surfaces. These wedges are often backed with additional sound-absorbing material. Ideally, none of the sound energy is reflected back into the test area.

Sound waves decay exponentially with distance into the sound-absorbing layers. Typically, it takes a minimum of one-quarter wavelength for the sound to decay to acceptable levels. At 20Hz the wavelength in air is

$$\lambda = \frac{c}{f} = \frac{1130}{20} = 56.5\,ft$$

so that one-quarter wavelength is about 14′ in air. Fortunately, sound slows down in the absorbing layer.

Assuming a 30% reduction in sound speed, the sound-absorbing material on each wall will still need to be at least 10′ thick if an anechoic chamber is to be effective down to the lower audible limit of 20Hz. This translates to minimum room dimensions of at least 20′ on a side. Allowing for more complete absorption and a reasonable work volume, the room dimensions in practice will be much larger. In order to keep the test volume to a reasonable size, many commercial anechoic facilities limit low-frequency extension to 50 or 100Hz. In some applications, such as machinery noise evaluation, even higher lower limits are common.

In a fully reverberant room, all sides are very hard and completely reflect incident acoustic energy. Fully reverberant rooms find application in a number of engineering and acoustical tests, but clearly such an environment is very hostile to loudspeaker testing. The arrival of direct energy from the loudspeaker, coupled with multiple arrivals of delayed energy from the reflecting walls at the test microphone, will cause severe response aberrations and, in particular, comb filtering of the measured response. (I will give an example of comb filtering shortly.)

The surfaces of a semireverberant room will have combinations of reflective and absorptive areas. Monitoring studios, theaters, and home listening rooms are examples of a semireverberant environment. Although not ideal, most of us must be content with a semireverberant room for loudspeaker testing.

Another way to classify an acoustic environment is in terms of the volume into which a driver or loudspeaker radiates. Two classifications we will use are full-space and half-space. A loudspeaker radiates into a full-space when its dimensions or the dimensions of its enclosure are small compared to the radiated wavelength, and when there are no nearby surfaces to reflect the sound. Full-space conditions can be attained in an anechoic chamber or by suspending the loudspeaker above the floor and far from all reflecting surfaces. It is often referred to as 4π radiation since the driver radiates into a full sphere, which is a solid angle of 4π steradians.

Half-space conditions are attained by baffling the loudspeaker on a very large wall, which effectively isolates front and rear radiation. It is also called 2π radiation, since the driver now radiates into a hemisphere or a solid angle of 2π steradians. Radiated sound pressure doubles under half-space loading. As a point of interest, it is common practice to rate driver sensitivity under half-space conditions.

4.3.2 TESTING LOUDSPEAKERS IN SEMIREVERBERANT ROOMS

Clearly, very few readers will have the room or resources to build an anechoic chamber, and fewer will have access to one. You can approximate anechoic conditions by placing the loudspeaker under test far from all reflecting surfaces. To get down to 20Hz, all reflecting surfaces should be at least one-half wavelength away, i.e., 28′.

This is easily accomplished with a 50′ crane or a 28′ tower. Another approach is to bury the loudspeaker in the ground, flush with the surface and at least 28′ from any reflecting surface other than the ground. This is known as half-space or half-plane testing, which I will say more about shortly.

The suggestions in the previous paragraph are not terribly practical and were offered somewhat with tongue-in-cheek. Most of you will have to use your listening room or perhaps a large open area in a basement or garage for loudspeaker testing. There is a widespread belief that you cannot make accurate response measurements in a typical listening room. However, it is not necessary to go to the extremes of the previous paragraph to get good data. By understanding room acoustics, using the right instrumentation and test signals, and—most importantly—using measurement techniques that complement or circumvent room limitations, it is possible to produce very good results.

Rectangular Room Acoustics: Let's talk a bit about room acoustics and, in particular, standing waves. The condition for establishing a standing wave in the one-dimensional case is illustrated in *Fig. 4.5*. A signal source placed on wall *a* launches an acoustic wave which travels to wall *b* and is reflected back toward *a* again. In general, the reflected wave will not be in-phase with the initial wave and will interfere with it, either constructively or destructively. However, when the total path length down and back equals one wavelength, the reflected wave will arrive back in-phase with the initial wave and a standing wave will be established.

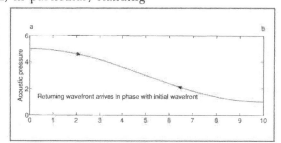

FIGURE 4.5a: Condition for a standing wave between walls a and b.

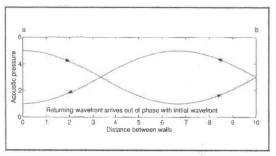

FIGURE 4.5b: One condition for no standing wave between walls a and b.

The lowest frequency standing wave will occur when the wavelength equals twice the wall spacing or, equivalently, when the wall spacing is one-half of a wavelength. Additionally, higher-frequency standing waves occur every time a reflected wave arrives back in-phase with the originating wave. This happens when the distance between the surfaces is an integer multiple of one half-wavelength. For the one-dimensional case we can determine the standing-wave frequency progression with the following formula:

$$f = \frac{c}{2} \times \frac{n}{L} \qquad [4.1a]$$

where:

c = speed of sound in m/s or ft/s
L = wall separation in meters or feet
n = is an integer 0, 1, 2, 3, etc.

c and L must be in consistent units. Use either Imperial units (feet and seconds) or SI units (meters and seconds). *Do not* mix units. With Imperial units equation 4.1a becomes:

$$f = \frac{1130}{2} \times \frac{n}{L} = 565\frac{n}{L} \qquad [4.1b]$$

Let's say the wall spacing is 18′. *Table 4.1* shows the first five standing-wave modal frequencies.

Standing-wave model patterns in rectangular rooms are much more complex. The strongest

TABLE 4.1
**MODAL FREQUENCIES FOR AN 18' WALL
SEPARATION**

N	F(HZ)
1	31.4
2	62.8
3	94.2
4	125.6
5	156.9

modes occur between two parallel walls. These modes are called *axial modes*. Reflections are also possible off-pairs of parallel walls. For instance, a wave launched from an end wall can hit a side wall, bounce to the opposite end wall, bounce again to the opposing side wall, and finally return to the first end wall.

Standing waves involving two pairs of parallel walls are called *tangential modes*. These tend to be weaker than the axial modes because they bounce off three walls and travel greater distances. The most complex modes are the *oblique modes* which involve reflections off all three pairs of parallel walls. As you might expect, these are the weakest.

Standing-wave modes for a rectangular room are computed with a generalization of Equation 4.1, first developed by Lord Rayleigh in 1896.

$$f = \frac{c}{2}\sqrt{\left(\frac{n}{L}\right)^2 + \left(\frac{l}{W}\right)^2 + \left(\frac{k}{H}\right)^2} \qquad [4.2]$$

where:

c = speed of sound in ft/s or m/s
L = length of room in feet or meters
W = width of room in feet or meters
H = height of room in feet or meters
n, l, and k are the integers 0, 1, 2, 3, 4, and so forth

At low frequencies, where wavelengths are comparable to room dimensions, standing-wave modes are widely and irregularly spaced in frequency, producing gross variations in loudspeaker/room response. At those frequencies where the room supports a standing wave, response will be amplified. Similarly, in the spaces between standing-wave modes, response will be depressed.

At higher frequencies, where wavelengths are much smaller than room dimensions, standing-wave distributions are very dense in frequency; that is, they are numerous and very close together. Under these conditions room support of loudspeaker frequency response is very smooth as long as speaker and microphone are far from reflecting surfaces. In technical terms, the low-frequency region is referred to as the *discrete standing wave* region because the standing waves are widely spaced and easily observed. The high-frequency region is called the *statistical standing wave* region, because the modes are so numerous and so closely spaced in frequency that they are better described by a statistical distribution than by direct enumeration. In typical rooms of 2000–2500ft³, the transition between the two regions occurs gradually somewhere between 200–400Hz.

An example will make this discussion clear. I used Equation 4.2 to compute the standing-wave frequency distribution for a typical listening room 8' H × 12.8' W × 18.6' L. These room dimensions are optimal in the sense that for a volume of 1,900ft³, they produce the most uniform distribution of modes. *Figure 4.6* shows the results of this computation for the first four octaves of room support. The presence of axial, tangential, and oblique modes is indicated by a "+" sign on the plot. The vertical position of the point indicates the relative strength of the mode. The axial

TABLE 4.2
**MODEL DENSITY FOR RECTANGULAR
ROOM EXAMPLE**

Octave (Hz)	Average Spacing (Hz) (All Modes)	Average Spacing (Hz) (Axial + Tangential)
30–60	10.1	10.1
60–120	3.3	3.3
120–240	1.3	1.5
240–480	1.1	1.0

modes are the least dense; the oblique modes the most dense.

There are 18 axial modes starting at 30.4Hz, the lowest frequency supported by this room. In the first octave between 30–60Hz there are only four modes. The average spacing between modes in the first octave is 10.1Hz. There are 108 tangential modes starting at 53.7Hz, and 216 oblique modes starting at 88.7Hz. *Table 4.2* lists the average spacing between modes for each of the first four octaves.

Table 4.2 lists modal density for all three modes, and also for just the axial and tangential modes. You might argue that the first two modes provide the strongest support for loudspeaker response relative to the weaker oblique modes. Each mode represents a room resonance with a resonance response curve much like a bandpass filter. The bandwidth of the response curve at the −3dB points is a measure of room Q at each frequency. The bandwidth depends on room absorption. Bandwidths are typically a few hertz, so when the spacing between modes falls below this range, room support of loudspeaker response will be uniform. For this example, room support can be considered uniform in the fourth octave.

Loudspeaker placement critically influences which room modes will be excited. All modes will

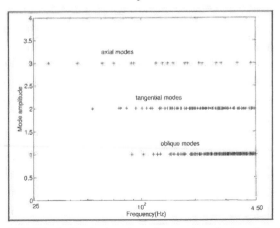

FIGURE 4.6: Distribution of standing-wave modes in rectangular room.

be excited if the source is placed at the intersection of three walls (in a corner at either the floor or the ceiling). Fewer modes will be excited if the source is placed at the intersection of two walls, and the fewest modes will be excited when the source is placed in the center of one surface. The effect of standing waves can be reduced by placing the speaker under test in the middle of the room far from reflecting surfaces.

Before leaving this topic, I want to dispel one popular myth. Nonparallel walls and sloping ceilings *do not* eliminate standing waves. The pressure distribution patterns become more irregular, but the number of modes will be about the same if the total volume is the same. Irregularly shaped rooms may have a more uniform distribution of modes and the nonparallel walls will reduce slap echo.

Reflecting Surfaces: We now know how to reduce the effect of standing waves, but we still have a problem with reflecting surfaces in our test room. Our goal is to measure the direct response from the loudspeaker. Reflected energy from the nearby walls, floor, and ceiling will arrive at the test microphone later than the direct wave. Depending on the path difference—therefore the phase difference—between the two arrivals, the later waves may add to or subtract from the direct wave. This alternate constructive and destructive interference is referred to as *comb filtering*.

Let's look at the effect of a single-reflecting surface. Typically, if the speaker under test is placed in the middle of the floor—far from reflecting walls— the first reflection will come from the floor. This condition is illustrated in *Fig. 4.7*. The driver to be tested and the test microphone are both at a height of one meter. The direct distance, d_1, from driver to microphone, is also one meter. The wave reflected by the floor travels the longer path, $2d_2$.

Whenever the distance $2d_2 - d_1$ is equal to an even multiple of a wavelength, the direct and reflected waves will add directly. Whenever this distance is an odd number of half wavelengths, the reflected wave will be 180° out-of-phase and subtract from the direct wave. At intermediate distance to wavelength ratios, there will be partial addition or subtraction. The reflected wave will be somewhat weaker than the direct wave, since it travels a longer distance. The phase angle of the reflected wave will lag the direct wave. The phase difference can be computed as follows:

$$\phi = 360\left(\frac{2d_2 - d_1}{\lambda}\right) = 360\left(\frac{(2d_2 - d_1)f}{c}\right) \text{degrees} \quad [4.3]$$

where λ is the wavelength. d_1, d_2, and λ must be in the same system of units. If we assume the driver is acting like a point source, sound pressure falls off inversely with distance. In this case, the ratio of reflected to direct-pressure waves is

$$\frac{P_R}{P_D} = \frac{d_1}{2d_2} \quad [4.4]$$

The distance, d_2, can be computed as the hypotenuse of a right triangle:

$$d_2 = \sqrt{h^2 + \left(\frac{d_1}{2}\right)^2} \quad [4.5]$$

Substituting the values from *Fig. 4.7* into the preceding equations we get:

$$d_2 = 1.118\text{m} \quad [4.6a]$$

$$\phi = 360\left(\frac{(2.236 - 1)f}{344}\right) = 1.294f \quad [4.6b]$$

$$\frac{P_R}{P_D} = \frac{1}{2.236} = 0.447 \quad [4.6c]$$

We can solve Equation 4.6b for the frequency of the first floor-induced response dip. At that frequency the phase angle will be 180°.

$$1.294f = 180$$

or

$$f = 180/1.294 = 139.1\text{Hz}$$

The first floor-induced peak will occur when ϕ equals 360°.

$$f = 360/1.294 = 278.2\text{Hz}$$

Assuming a perfectly flat driver response, the response caused by the floor reflection is shown in *Fig. 4.8*. When plotted on a linear frequency scale, we see

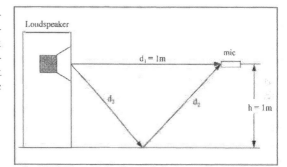

FIGURE 4.7: Floor reflection geometry.

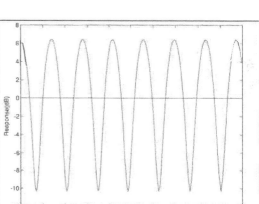

FIGURE 4.8: Comb-filter response caused by floor reflection.

that it consists of alternating dips and peaks resembling the teeth of a comb. The floor reflection produces a comb-filter response. The peak-to-peak (pk-pk) variation and the spacing of the dips are a function of the microphone/driver height, h, and the distance, d_1.

Increasing h or decreasing d_1 will move the first dip lower in frequency and spread the spacing of the successive dips and peaks. If you increase h too much, however, you will encounter the ceiling reflection. You also cannot get too close to the driver or you will get into its near-field response, which differs from the far-field response. I will discuss the near-field/far-field issue shortly.

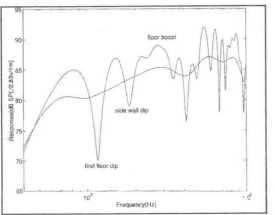

FIGURE 4.9: Response of 180mm driver in reverberant room.

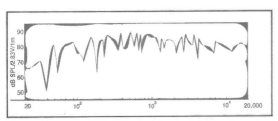

FIGURE 4.10a: Two-way system response in reverberant room.

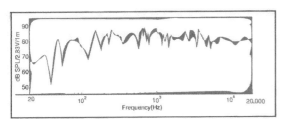

FIGURE 4.10b: Response with two fiberglass batts on floor.

Figure 4.9 shows the frequency response of a 180mm mid-bass driver taken in a highly reverberant room. The driver and microphone are 1.1m off the floor, and the microphone is positioned 1m from the driver baffle. The first-floor dip occurs at 121Hz, followed by a side-wall dip at about 180Hz, and the first-floor boost at about 240Hz. The peaks and dips become more numerous after that, and their exact cause is difficult to determine.

The smooth curve lying wholly within the frequency-response curve is the driver's anechoic response. Thick fiberglass batts placed on the floor between the DUT and the microphone can reduce some response irregularities caused by floor bounce. *Figure 4.10a* shows the response of a free-standing small two-way monitor loudspeaker made with a high-resolution sine-wave sweep in a fairly reverberant room. *Figure 4.10b* shows

and the popular audio press have spread a great deal of misinformation about diffraction. We commonly see statements like: "rounded corners eliminate diffraction," "spherical enclosures eliminate diffraction," and "cylindrical enclosures eliminate diffraction." You can often mitigate diffraction effects with appropriate geometry and absorbing materials, but it is almost impossible to eliminate it.

Having said that, there is one sure way to eliminate diffraction in a loudspeaker. All you need is a very large wall with length and width at least ten times the wavelength of the lowest frequency you want to reproduce ($560' \times 560'$ for 20Hz). Mount your loudspeaker in the center of this wall, flush with the wall surface. Viola! No diffraction!

What is *acoustic diffraction*? Here are two definitions:

1. Diffraction is the distortion of a wave front caused by the presence of an obstacle in the sound field.[2]

2. Diffraction is the change in direction of propagation of a wave front due to the presence of an obstacle or discontinuity.[3]

Acoustic diffraction theory is very complex because it involves the solution of partial differential equations with obscure boundary conditions. In the past, only very simple geometries could be analyzed. However, with the advent of powerful computers and finite element and boundary element analysis techniques, a greater understanding of the physics of diffraction has evolved. Fortunately, the two manifestations of diffraction that most directly impact loudspeaker response and testing are relatively easy to describe. They are low-frequency *spreading loss* and *edge diffraction*. Let's look at spreading loss first.

Spreading loss: A typical loudspeaker will have its drivers mounted on a rectangular baffle. Compared to a wavelength, at very low frequencies, baffle dimensions will be small. In this low frequency range, radiated sound easily wraps around the enclosure, making the loudspeaker omni-directional. As frequency increases and baffle dimensions become comparable to the wavelength of the radiated sound, the baffle begins to act like a reflecting surface, increasing SPL in the forward direction. This is what typically happens

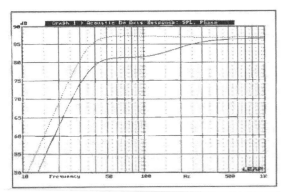

FIGURE 4.11: Half-space versus anechoic response (dotted = half-space, solid = free-standing anechoic).

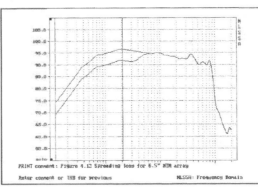

FIGURE 4.12: Spreading loss for 6.5″ MTM array.

the same system with two 6″ fiberglass batts placed on the floor. The deep notch just below 200Hz is eliminated in the second plot, and response variations above 200Hz are somewhat reduced, but reflections from the walls and ceiling are still evident.

Baffles and Diffraction: Marketing departments

when a loudspeaker is placed in a listening room. We have seen this phenomenon before in connection with the discussion on microphones.

Interestingly, the process is the same whether it is a microphone surface intercepting an impinging wave or a baffle reflecting a generated wave. At very high frequencies relative to baffle dimensions, just about all sound is radiating in the forward direction. Thus, over the full frequency range the system transitions from full-space to half-space radiation and the *on-axis* SPL doubles; that is, increases by 6dB.

Figure 4.11 shows the computer-simulated response of an ideal 180mm driver mounted on a very large wall (half-space radiation) and compares this with the same driver mounted on a $10'' \times 14.5''$ rectangular baffle. In the latter case response begins to fall off with decreasing frequency above 1kHz, but most of the response drop occurs over the two octaves from 100–400Hz. At 70Hz, response is down by 6dB relative to the 1kHz value. In practice, room modes, surface reflections, and driver-response variations may partially mask spreading loss.

Figure 4.12 contains frequency-response plots of a pair of 6.5″ mid/bass drivers arranged in an MTM array, spaced 11″ apart center-to-center on a narrow baffle. Both half-space and anechoic or free-field results are shown. The spreading loss begins around 700Hz. However, this loss is partially counteracted by the rising response of the driver pair as seen in the half-space curve. The maximum loss of 4.7dB relative to the half-space response occurs at 200Hz.

Edge Diffraction: Edge diffraction is best demonstrated with what is perhaps a worst-case example. *Figure 4.13* shows the geometry. A point source is mounted in the center of a circular disk 2R inches diameter. A conceptual picture of the edge-diffraction process is shown in *Fig. 4.14*. The source is driven with a pure tone producing a hemispherical wave front progressing outward along the disk surface. When the wave reaches the edge of the disk, it is suddenly forced to expand into a much larger volume. The pressure at this point must drop.

The original wave continues to expand outward, wrapping around the disk and diffracting to the rear with no change in phase. The pressure drop at disk edge, however, causes a second wave to be launched at the disk edge traveling in the forward direction. The *phase* of this wave is *reversed* relative to the original wave. One way to view this is to consider the drop in pressure caused by the generation of a second wave at the disk's edge with opposite polarity to the original or incident wave.[3,4]

The forward propagating diffracted wave will interfere with the original wave. Referring to *Fig. 4.13*, the forwardly diffracted wave travels a distance $R + d_2$, where

$$d_2 = \sqrt{R^2 + d_1^2}$$

Because of the phase reversal, the combined response along the tweeter's principal axis will be a *minimum* whenever the path length difference,

$$\Delta = R + d_2 - d_1,$$

is a whole number of wavelengths. That is, when

$$\Delta = n\lambda, \ n = 1, 2, 3, \text{ etc.}$$

It will be a *maximum* when Δ equals an odd number of half-wavelengths or when

$$\Delta = (2n - 1)\lambda/2, \ n = 1, 2, 3, \text{ etc.}$$

Recalling that $\lambda = c/f$, we can solve for the minimum and maximum frequencies.

$$f_{min} = \frac{nc}{\Delta}$$

and

$$f_{max} = \frac{(2n - 1)c}{2\Delta}$$

Consider a 20″ diameter disk with a microphone placed at a distance of 40″ from the disk. For this example, Δ is computed as follows:

$$d_2 = \sqrt{10^2 + 40^2} = 41.2''$$

and

$$\Delta = 10 + 41.2 - 40 = 11.2'' = 28.5\text{cm} = 0.285\text{m}$$

The first minimum will occur at

$$f_{min} = c/\Delta = 344/0.285 = 1207\text{Hz}$$

Subsequent mini-ma will occur at 2414Hz, 3621Hz, 4828Hz and so forth, while max-ima will occur at 603Hz, 1809Hz, 3015Hz, 4221Hz, etc.

I made a mock-up of this example by mounting a 19mm soft-dome tweeter in the center of a 20″ diameter, ¾″ particleboard disk. *Figure 4.15* shows the results of response measurements made with a microphone placed 40″ away from the tweeter on the tweeter axis. Response dips are clearly evident at 1.2k, 2.4k, 3.6k and 4.8kHz, in excellent agreement with the previous calculations. Local peaks can be seen at roughly 1.8k, 3.0k, 4.2k and 5.4kHz. The first peak at 600Hz is barely visible because it is greatly suppressed by the rapidly falling tweeter response in that frequency region.

Incidentally, the disk circumference had a full ¾″ rounded edge. So much for rounded edges elimi-

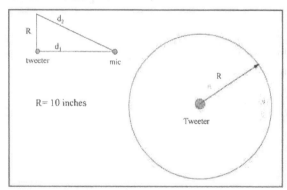

FIGURE 4.13: Diffraction example: tweeter on circular disk.

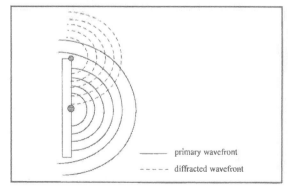

FIGURE 4.14: Conceptual view of edge diffraction.

nating diffraction. Diffraction is caused by the rapid expansion of the wave front into the larger volume beyond the disk's edge. For rounded edges to be effective in reducing diffraction, the edge radius must be comparable to a wavelength. For a ¾″ round this corresponds to a frequency of 18kHz.

One point not yet made is that diffraction effects are highly directional. Part of this has to do with geometry and part has to do with the nature of diffraction physics. The dotted curve in *Fig. 4.15* represents the tweeter response 30° off-axis. The plot is much smoother than the on-axis result. Each incremental distance along the disk circumference acts like an independent secondary source. In the on-axis case the path length from each incremental length along the disk's edge to the microphone is the same so that all diffracted energy reaches the microphone with the same phase, and thus, has the maximum effect.

In the off-axis case, some parts of the disk edge are closer to the microphone, while others are farther away. The diffracted energy now arrives with a broad range of phase angles, some subtracting and some adding to the direct wave. Surprisingly, diffraction effects disappear entirely at 90° off-axis.

Clearly, centering a driver on a circular disk is a poor idea. Mounting the driver off-center will greatly reduce on-axis diffraction effects. Rectangular baffles are superior to circular baffles even when the driver is centered, because the distance to the baffle edges is not constant.

Figure 4.16 shows the response of a 28mm soft-dome tweeter mounted in the center of a 2′ × 2′ baffle. The only significant diffraction dip is at about 800Hz. There are some additional smaller diffraction-induced response ripples in the 2–5kHz

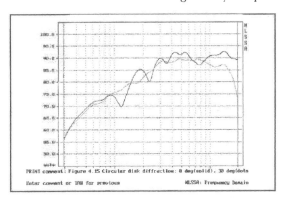

FIGURE 4.15: Circular disk diffraction: 0° (solid), 30° (dots).

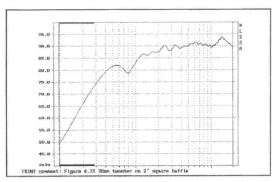

FIGURE 4.16: 28mm tweeter on 2′ × 2′ baffle at 1m.

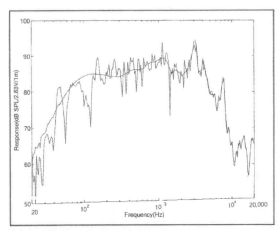

FIGURE 4.17: 6.5″ driver: swept sine wave versus anechoic response.

range. Diffraction response ripples can be largely eliminated by mounting the tweeter on this baffle such that no two edges are the same distance from the tweeter.

There is one final source of diffraction error in loudspeaker testing: the microphone mount. Microphones for loudspeaker testing have a small diameter to present minimum frontal area to the measured wave front. Bulky mounting fixtures, however, provide reflective surfaces that can produce frequency-response combing. Because these surfaces are relatively near the microphone, the combing response tends to occur at higher frequencies and is easily misinterpreted as being produced by the DUT. This is especially true when testing tweeters. Ideally, any structural members supporting a microphone should have a smaller diameter than the microphone itself. Covering supporting booms with absorptive foam is also a good idea.

I have spent a great deal of time describing the errors introduced into acoustic loudspeaker measurements by the environment and baffle geometry so that you will be able to recognize these effects when you see them and distinguish them from driver response characteristics. In the next two sections we will examine relatively simple measurement techniques and test signals that can greatly reduce these effects and give you a truer picture of loudspeaker frequency response.

4.4 LOUDSPEAKER MEASUREMENT TECHNIQUES AND TEST SIGNALS

In this section we will examine three measurement techniques, detail their advantages and disadvantages, and the frequency range over which each technique is best suited. The three techniques are:

1. Far-field measurements
2. Near-field measurements
3. Ground-plane measurements

We will also look at different test signals—specifically, pink noise and warble tones—to see how they can be combined with the measurement techniques, providing very good results.

For a preview of where we are going in this section, look at *Fig. 4.17*. It is a plot of the stepped sine-wave response of a 6.5″ mid/bass driver mounted on the 10″ × 14″ front baffle of a 10 ltr closed-box enclosure. The response was taken in the center of a large, semireverberant room with the driver mounted on a stand at a height of 1m. The microphone was also placed at a height of 1m, 1m in front of the DUT. The sine-wave source was stepped in 1/24th octave increments from 20Hz–20kHz to provide a very high resolution response plot.

Also shown on this plot is the true anechoic response of the DUT. This is not a particularly smooth driver. It was deliberately chosen for its response wiggles, which we will learn to differentiate from room-induced and diffraction effects in the following sections.

Referring to *Fig. 4.17*, notice that in the region above 300Hz, the mean or average of the stepped sine-wave response tends to follow the anechoic response. This is the statistical standing-wave region where response data is dominated by reflections rather than standing waves. You can see that the response caused by these reflections tends to cluster around the true response.

Clearly, some type of averaging or smoothing of the high-resolution sine-wave response should yield a close approximation to the anechoic response above 300Hz. If the data is available in digital form, the smoothing can be done numerically in a computer. This is what happens with the PC-based systems, which will be discussed in Chapter 7. In this chapter we will see how pink noise and warble tones can be used to do the smoothing in the analog domain.

Below 300Hz we can see very large dips and peaks in the stepped sine-wave response. In this region, measured response is dominated by standing-wave effects. Smoothing will not help here. Near-field measurements will be required. In the near-field technique, the microphone is placed very close to the driver diaphragm to swamp out diffraction and room effects. Within its range of validity, near-field measurements produce the true half-space response.

In a ground-plane measurement, the DUT is placed directly on a hard surface, such as a concrete floor or paved driveway. This eliminates the floor reflection. If the DUT is also far from any other reflecting surfaces, such as a large basement floor or a parking lot, the ground-plane technique will yield the DUT's true anechoic response raised by 6dB.

**TABLE 4.3
PRESSURE RATIOS VERSUS DISTANCE
FOR A 6.5″ DRIVER**

Distance Ratio	dB Difference
12/6	−5.75
24/12	−5.92
48/24	−6.03

4.4.1 THE FAR-FIELD

In a driver's far field, sound-pressure level falls off inversely with distance so that the driver looks like a point source. That is, sound-pressure level drops 6dB with each doubling of distance, and sound radiated from each elemental area of the diaphragm arrives coherently at the listening or measurement position. We generally listen to drivers in the far field. Line sources, such as tall, narrow ribbon drivers, and tweeter line arrays are exceptions to this definition. For these loudspeakers sound-pressure level falls by 3dB for each doubling of distance. (Of course, if you go out far enough, typically 10–20 times the array length, even the line array will start to look like a point source.)

As a general rule-of-thumb, you should be in the far field at a distance equal to three to five times the driver diaphragm diameter. For all tweeter and mid-range drivers and for most woofers, you are well into the far field at one meter. Continuing

with our mid/bass driver example of *Fig. 4.17*, *Fig. 4.18* shows the quasi-anechoic pressure response of the 6.5″ driver at distances from its baffle of 6″, 12″, 24″, and 48″. This data was taken with the MLSSA PC-based acoustic measurement system, which will be described in Chapter 7. *Table 4.3* lists the pressure drop as a function of the distance ratios.

The data in *Table 4.3* is slightly in error because the distances were measured most conveniently from the baffle board and do not quite coincide with the driver acoustic center, which is about 1.2″ behind the baffle board. Nevertheless, we are into the driver's far field somewhere just beyond 24″ from the baffle board, or a-bout five times the effective diaphragm diameter for this driver. Notice that the shape of the driver's response curve changes subtly from 6–24″ out, especially in the 1kHz–3kHz range. The 24″ and 48″ curves, however, are essentially identical in shape, which is another indication that we are finally in the far field.

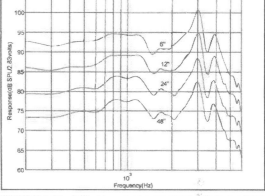

FIGURE 4.18: 6.5″ driver response versus distance from baffle.

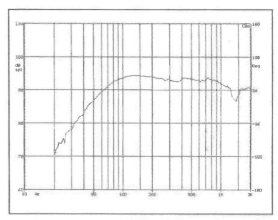

FIGURE 4.19: Near-field response of 6.5″ driver.

4.4.2 THE NEAR-FIELD

The near-field technique is used to overcome the effects of low-frequency standing waves. In this technique, the microphone is placed very close to the driver diaphragm to swamp out diffraction and room effects. At low frequencies where the driver diaphragm behaves like a rigid piston, the measured near-field response is directly proportional to the far-field response and independent of the environment into which the driver radiates. D.B. Keele describes this technique in his excellent paper.[5] I will summarize the approach and its limitations here.

For the near-field technique to work properly, the microphone should be placed as near to the center of the diaphragm as possible. Keele shows that a microphone distance less than 0.11 times the diaphragm effective radius results in measurement errors of less than 1dB. The 6.5″ driver we have been examining has an effective cone diameter of 5″ or an effective radius of 2.5″. For this driver, the microphone should be placed within 0.275″ of the driver dust cap.

At higher frequencies, pressure waves from various areas of the diaphragm may arrive at the microphone out-of-phase, causing near-field response cancellations not observed at normal listening distances. For this reason there is a practical upper limit to the near-field technique given in terms of driver diaphragm diameter. For a driver mounted in an infinite baffle, the limit is

$$f_{MAX} = \frac{4311}{D} \qquad [4.4.1a]$$

where f_{MAX} is in hertz and the driver diameter, D, is in inches, or

$$f_{MAX} = \frac{10,950}{D} \qquad [4.4.1b]$$

where D is now in centimeters. For closed-box or ported systems with finite baffles, this limit may be slightly lower, but Equation 4.4.1 will put you in the ball park. For our 6.5″ driver we have:

$$f_{MAX} = \frac{4311}{5} = 862\,Hz$$

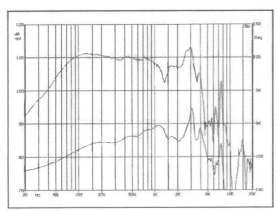

FIGURE 4.20: Quasi-anechoic and near-field responses.

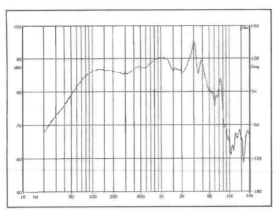

FIGURE 4.21: Merged near- and far-field responses.

Figure 4.19 shows the stepped sine wave near-field response of our 6.5″ mid/bass driver. The microphone was placed within 0.25″ of the driver dust cap. Although the plot is valid only up to 862Hz, the sine-wave frequency was allowed to range from 20Hz–2kHz in 1/12 octave steps in this test. This data was collected with the CLIO PC-based acoustic analysis system, which will be discussed further in Chapter 7.

Notice that the response is entirely free of room effects. This is the actual measured response. No smoothing was applied to the data. The plot is characteristic of a slightly underdamped closed-box system. The system Q is 1.1.

One final step remains in the near-field technique: converting the near-field measurement to an equivalent far-field response. For a driver radiating omnidirectionally into a 4π space, the conversion in dB is[6]

$$R_{FF} = R_{NF} - 20\log_{10}\left(\frac{4d}{r}\right) dB \qquad [4.4.2a]$$

For a driver radiating hemispherically into a 2π space the conversion is

$$R_{FF} = R_{NF} - 20\log_{10}\left(\frac{2d}{r}\right) dB \qquad [4.4.2b]$$

R_{NF} is the near-field response and R_{FF} is the far-field response at a distance, d, from a driver with an effective diaphragm radius of r. Both r and d must be consistent in terms of length units. The second term of the right-hand side of each equation above—the correction term—needs to be computed only once and then applied to every data point. Although these conversion formulas are useful, in practice, the conversion is usually done visually by properly merging near- and far-field responses as the example in the next section will show.

4.4.3 MERGING NEAR- AND FAR-FIELD RESPONSES

Figure 4.20 shows the far-field and near-field responses of our 6.5″ driver on a common frequency scale. This data was collected with the CLIO system in the MLS analysis mode. Although this chapter deals with simple analog techniques, I used the PC-based system to collect far-field data free of room reflections for this example. Here the presence of room reflections would just confuse the main thrust of the discussion.

The lower plot is the far-field response. It represents the first 4.5ms of the driver impulse response, and is free of room reflections and valid above 220Hz. Notice that this curve falls rapidly below 220Hz because the gated data segment contains no signal energy with periods longer than 4.5ms. The far-field data was taken at a distance of 1m with 2.83V input, and thus, represents the true driver sensitivity.

The upper curve in *Fig. 4.20* shows the near-field response taken 0.25″ in front of the dust cap. The first 50ms of the near-field impulse response was used to compute the curve which makes it valid down to 20Hz. Comparing *Figs. 4.19* and *4.20* and accounting for the difference in frequency scales, you can see that the near-field data taken with stepped sine waves is in excellent agreement with the MLS technique.

This is the beauty of the near-field technique. Near-field measurements are independent of the measurement technique. They have a very high signal-to-noise ratio due to the closeness of the microphone, providing excellent rejection of reflections and random background noise.

The near-field data is valid below 862Hz, while the far-field data is valid above 220Hz. Clearly, the two graphs should be joined somewhere in the 220–862Hz range. There are several possible choices. Regardless of the point you choose, however, the near-field response should always be brought into coincidence with the far-field response, since the latter represents the true driver sensitivity. If the graphs are joined at too high a frequency, the full spreading-loss drop due to finite baffle size may be missed.

From *Fig. 4.20* we see that the far-field response levels off just above 300Hz and is flat to the lower valid limit of 220Hz. The near-field response is also relatively flat in the 250–300Hz range. 280Hz looks like a good point to join the two graphs.

Figure 4.21 shows the merged response where the two plots have been joined at 280Hz and represents the full-range far-field anechoic driver response. This is the same graph used in *Fig. 4.17.* Note that the far-field response determines whether the final merged graph is an anechoic or half-space response. Since our far-field graph included the effect of spreading loss, it is an anechoic plot.

4.4.4 GROUND-PLANE MEASUREMENTS

So far our emphasis has been on trying to remove or reduce the effects of the acoustic environment on our measurements. The near-field technique of the previous section is an example of that approach. Reflections have been particularly troublesome. We now follow the old adage, "if you can't beat 'em, join 'em." In the ground-plane technique, we deliberately introduce a strong reflection, but the effect of this reflection is entirely predictable and beneficial.[7]

Figure 4.22a illustrates the ground-plane technique. The DUT is placed on a smooth, flat, rigid surface, and the microphone is placed flush with the surface at a suitable measuring distance. The strong reflection at the surface produced by the DUT can be thought of as a reflection-free driver on the surface combined with a second loudspeaker placed just under the surface. This second loudspeaker is called the *image of the DUT.*

The ground-plane setup is equivalent to the measurement scenario shown in *Fig. 4.22b.* Here we see two identical DUTs with the microphone placed along the centerline between them. Ideally, the DUT should be tilted slightly so that the microphone is on-axis. Referring to *Fig. 4.24,* the tilt angle may be computed as:

$$\vartheta = \arctan\left(\frac{H}{d}\right) \qquad [4.4.3]$$

In practice, tilting the DUT may not be required, since θ is often quite small, so that a slight off-axis measurement will not differ greatly from the true on-axis response.

Several comments are in order concerning this technique. First, the presence of a virtual second DUT increases the on-axis SPL at any microphone distance by 6dB relative to the SPL of a single driver. The microphone should be placed in the far-field at three to five times the largest dimension of the source. When testing multi-driver systems the source extent includes the span of all drivers on the baffle. As long as far-field conditions are met, it is common practice to place the microphone 2m

from the DUT. Because of the 6dB boost, this placement will yield the normal one meter sensitivity.

The ground-plane technique simulates two sources in free space positioned in mirror image along the measurement axis. For this reason, the effective baffle area is twice as large and the shape is different from that of the single system. There is also some mutual coupling between the DUT and its image. As a result, frequency response will be somewhat different from the free-space measurement.

This can be eliminated with a rather unusual application of the ground-plane technique. If the DUT is placed above the ground plane (GP) by 2m, its image will be below the GP by 2m, effectively uncoupling the DUT and its image. The practical advantages in placing the DUT on the ground, however, will generally make the small measurement error acceptable.

Figure 4.23 is a computer simulation showing the differences you can expect between anechoic or free-field, half-space and ground-plane measurements. This is the same simulation first shown in *Fig. 4.11* with the addition of the ground-plane effect. Notice the small difference in the low-frequency shape of the half-space and ground-plane curves.

The microphone must be placed on the centerline between the DUT and its image. Because we have, in effect, two sources, any small offset from the centerline will mean that the distances from each source to the microphone are different. A dip in the floor might cause such an offset. At higher frequencies this small difference can lead to phase differences between the two signals and partial cancellation. The height offset will lead to an upper frequency limit for a specified accuracy of the

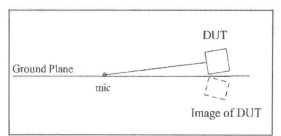

FIGURE 4.22a: Ground-plane measurement technique.

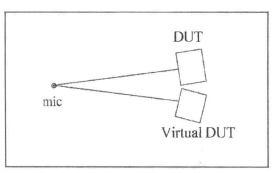

FIGURE 4.22b: Equivalent model for ground-plane measurement.

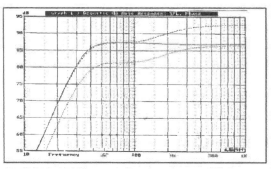

FIGURE 4.23: Anechoic, half-space, and ground-plane responses: ground plane (dashed), half space (solid), anechoic (dotted).

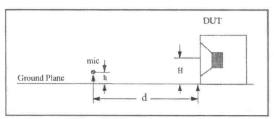

FIGURE 4.24: Illustrating height offset in ground-plane measurement.

ground-plane measurement.[6] Referring to *Fig. 4.24*, for a 1dB error:

$$f_{MAX} = \frac{25.7 \times 10^4}{Hh} \qquad [4.4.4a]$$

and for a 3dB error:

$$f_{MAX} = \frac{42.9 \times 10^4}{Hh} \qquad [4.4.4b]$$

H, the height of the DUT center, and h, the microphone height, are in centimeters.

From the previous paragraph we see that frequency-response errors occur only if the microphone is moved off the ground plane in the *vertical* direction. Off-axis measurements can be taken by moving the microphone in an arc on the ground plane. If the enclosure containing the DUT is laying on its side, vertical polar response can be determined by placing the microphone at various points on the ground plane, making sure that the distance to the DUT is kept constant. Rotating the cabinet 90° will permit measurement of the DUT's polar

times the DUT to microphone spacing. This will ensure that any errors caused by reflection are at least 20dB down, relative to the main arrival.

I measured ground-plane and conventional 1m stepped sine-wave responses on a 180mm mid/bass driver and mounted the driver in a 15 ltr box with baffle dimensions of 14″ × 10″. A convenient parking lot was not available, so I took these measurements on the smooth concrete floor of a large, fairly reverberant basement. For the conventional measurement, the DUT was stand-mounted at a height of 40″ with the microphone placed 1m from the baffle board. I placed 18″ of fiberglass batting on the floor between the DUT and the microphone to reduce floor bounce.

For the ground-plane measurement, the DUT was on the center of the floor, the microphone was also on the floor at a distance of 1m (39.37″), and the driver center was 5″ above the floor, requiring that the enclosure be tilted down by

$$\vartheta = \arctan\left(\frac{H}{d}\right) = \arctan\left(\frac{5}{39.37}\right) \doteq 7.2°$$

to place the microphone along the driver's central axis. (For those of you adverse to computing arc tangents, the box tilt can be determined as the ratio of the rise to the run. This ratio is approximately 5:40 or 1:8. To get the correct angle just move back 8″ from the front baffle and prop up the box one inch.) The nearest hard reflecting walls were 14′ away putting these reflections down about 18.6dB relative to the direct arrival. The ceiling height was 9′, but the rafters and wooden flooring break up and attenuate ceiling reflections substantially relative to those from the hard walls.

The results of these measurements are compared in *Fig. 4.25a*. The one meter data is offset by an additional 10dB for clarity. For both plots the statistical standing-wave region begins around 300Hz, but the pk-pk variations in the GP data above this frequency are smaller.

Below 300Hz, the one meter data is seriously contaminated by floor bounce and standing-wave effects. Contrast this with the GP data which shows only slightly larger variation than the data above

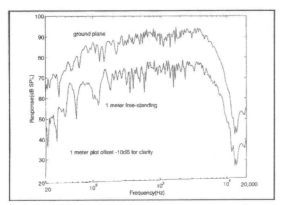

FIGURE 4.25a: Ground plane and 1m response of 180mm driver.

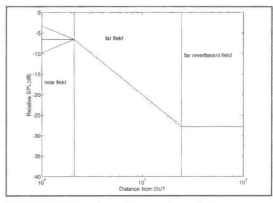

FIGURE 4.26: Near, far, and far reverberant fields.

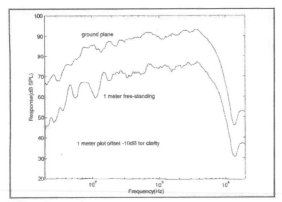

FIGURE 4.25b: 1/3 octave smoothed response of 180mm driver.

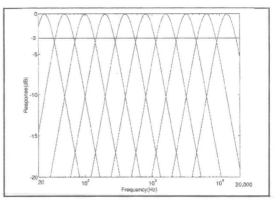

FIGURE 4.27: Octave-spaced bandpass filters on log frequency scale.

response in the horizontal plane.

One final source of error in the ground-plane technique is the presence of large obstructions on the GP which could cause reflections. Ideally, large obstructions in any direction should be at least five times farther away from the DUT than the microphone. The total path length of the reflection from the DUT to the microphone will then be ten

300Hz. The effect of standing waves and floor reflections is largely missing in the GP data. This is just the result we wanted.

Figure 4.25b shows a 1/3 octave smoothed version of the plots in *Fig. 4.25a*. This is actually a warble-tone response, which will be discussed further in Section 4.5. Clearly, smoothing of the low-frequency GP data is much more successful than the smoothing of one meter data and gives us a truer picture of the driver's low-frequency response. The average difference between the two curves is 6.5dB which is very close to the 6dB theoretical value. The difference is due to the contaminating effects of reflections and standing waves.

As a final point in this comparison, notice that the curves differ subtly in the range between 300Hz–4kHz. This is due to the baffle geometry differences discussed earlier. Although originally proposed for large open areas, the ground-plane technique is fairly effective even in somewhat confined spaces, as this example shows.

One final point before we move on to the topic of analog signal smoothing in the next section. *Fig. 4.26* is a conceptual plot of SPL as a function of distance from the DUT. There are three distinct regions—the near field, the far field, and the far reverberant field.

After a sufficient length of time in a room with reflective surfaces, a large number of generated reflections mix, producing a diffuse sound field. This is the *far reverberant field*. We want to measure the DUT's far-field response, but if we move too far away, our measurement will be seriously contaminated with the multiple reflections that make up the reverberant field. Thus, a critical balance exists in the measurement process. We must be out far enough to be in the far field, but not farther if we are to obtain the best signal-to-noise ratio.

In addition to the far reverberant field caused by our test signal, high-level, low-frequency background noise produced by HVAC equipment is often present in the test area. The impact of this noise can be reduced or eliminated with an "optional" high-pass filter (see *Fig. 4.1*). A 12 or 18dB/octave slope with a 200–300Hz breakpoint will have minimal effect on far-field measurements. The high-pass filter can be switched out for near-field measurements below 200Hz which are not effected by HVAC noise.

4.5 ANALOG SMOOTHING OF FREQUENCY RESPONSE CURVES

A bank of bandpass filters with appropriately spaced center frequencies can form the basis of a good audio analyzer. *Figure 4.27* shows the frequency response of ten bandpass filters with center fre-

TABLE 4.4	
RTA FILTER CENTER FREQUENCIES (HZ)	
20	63
25	80
32	100
40	125
50	160

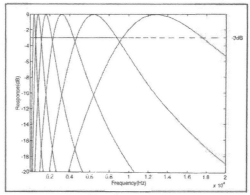

FIGURE 4.28: Octave-spaced bandpass filters on linear frequency scale.

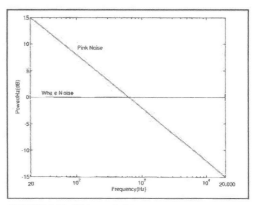

FIGURE 4.29: White and pink noise power spectra.

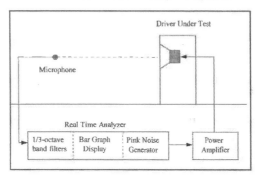

FIGURE 4.30: Setup for RTA measurement of loudspeaker frequency response.

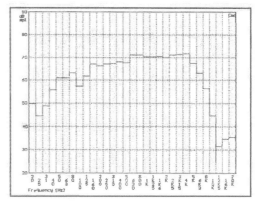

FIGURE 4.31: 1m RTA response of 180mm driver.

quencies spaced one octave apart starting at 25Hz and ending at 12.8kHz. If we can find a test signal such that the RMS output of each filter is a measure of the *average* SPL in the band it covers, this output will represent the smoothed response of our

65

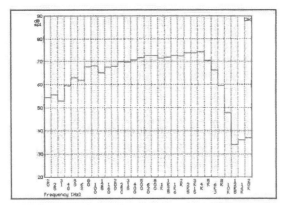

FIGURE 4.32: 180mm driver ground plane RTA plot.

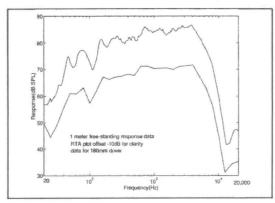

FIGURE 4.33: Comparison of RTA and smoothed sine wave responses.

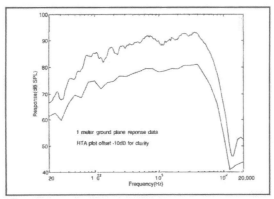

FIGURE 4.34: Comparison of GP RTA and smoothed sine wave responses.

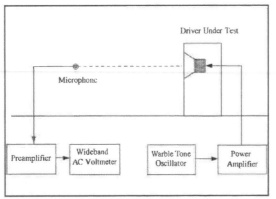

FIGURE 4.35: Warble-tone frequency response measurement setup.

66

DUT over that band.

4.5.1 BANDPASS FILTERS, PINK NOISE, AND THE RTA

Two signals that come to mind are *white noise* and *pink noise*. Both are wideband signals containing all frequencies within the audio band. They are easily generated with current IC technology. To see which signal is appropriate we must examine *bandpass filter* response in more detail. Referring to *Fig. 4.27*, all filters have the same Q so that all response curves have the same shape when plotted on a log-frequency scale. In particular, the −3dB bandwidths *appear* to be identical on this scale.

Q of a second-order bandpass filter is defined by the following equation:

$$Q = \frac{f_0}{f_2 - f_1} \qquad [4.5.1]$$

where

f_0 = bandpass filter center frequency
f_2 = upper −3dB frequency
f_1 = lower −3dB frequency

If we invert this equation we get:

$$\frac{f_2 - f_1}{f_0} = \frac{1}{Q}$$

In order to keep a constant Q, the quantity $f_2 - f_1$ must double for each doubling of the center frequency, f_0. When plotted on a *linear* frequency scale as shown in *Fig 4.28*, we indeed see that the bandwidth defined by the −3dB points doubles for each doubling of the center frequency. If Q is held constant, these bandpass filters will have constant *percentage* bandwidth. Therefore, we need a test signal with constant percentage bandwidth power. Since the actual bandwidth doubles for each octave increase of the center frequency, the wideband noise signal power must decrease by half or 3dB per octave to maintain the same RMS value at each filter output. But this is just the definition of pink noise.

Pink- and white-noise power spectra are compared in *Fig. 4.29*. The white-noise spectrum is flat over frequency. For any fixed absolute frequency range, $f_2 - f_1$, the power in the white-noise signal over that range is constant. For pink noise, however, the power over any fixed frequency *ratio* is constant. For the same total power over the entire 20Hz–20kHz range, the pink-noise spectrum must begin at a higher level and then fall below the white-noise level. The crossover point is at 632Hz.

The frequency resolution provided by octave-band filters is a bit too crude to be useful. One-third-octave (actually 1/10-decade) band filters provide higher resolution with acceptable averaging capability. The typical *real-time analyzer* (RTA) comprises a pink-noise generator, a bank of one-third-octave band filters, and a bar graph display. The setup for RTA-based loudspeaker response measurement is illustrated in *Fig. 4.30*.

The standard RTA has 31 1/3-octave band filters spaced in the frequency ratio:

$$R_{RTA} = 10^{0.1} = 1.259$$

The filters cover three decades from 20Hz to 20kHz with ten filters per decade and one final filter with a 20kHz center frequency. *Table 4.4* lists filter center frequencies for the first decade. The center frequencies of the second and third decades are just ten and 100 times the first, respectively.

Filter skirts for attenuating out-of-band signals may be 6, 12, 18, 24, or 36dB/octave. The shallower slopes are better for loudspeaker-response testing.

Analog RTAs are commercially available starting at around $1,000 US. Laboratory-grade RTAs can be considerably more expensive. LinearX makes a digitally controlled PC-based analog RTA called pcRTA, which I reviewed in *SB 6/95*. Most of the PC-based audio analysis systems discussed in Chapter 7 mechanize a fully digital RTA mode. An excellent pink/white noise generator kit is available from Old Colony Sound Lab, but this still leaves the home builder with the problem of designing and building the bank of 31 1/3-octave filters.

Because pink noise is a random process, and it takes a finite amount of time for standing waves and reverberation to be established in the test environment, there can be considerable sample-to-sample variation in measured 1/3-octave band responses. This is especially true for the lowest frequency bands. Most RTAs provide the capability to average several measurement sequences.

For greatest accuracy the output voltages from the filters should be measured with true RMS detectors rather than average-responding rectifiers. This is a further complication for the home experimenter who tries to build his own analog RTA, as the more commonly available average-responding AC voltmeter will be almost useless in the lowest-frequency bands.

Let's look at some RTA results obtained with the CLIO PC-based acoustic analysis system operating in the RTA mode. *Figure 4.31* shows the one meter, 1/3-octave response of the 180mm driver first discussed in Section 4.4. Notice that the graph has a stepped or staircase shape with a straight line spanning each 1/3-octave band at the measured SPL. This plot should be compared with *Figs. 4.25a* and *b*. We see that the general shape of the free-standing, one-meter response curve has been reproduced by the RTA. We have, however, lost considerable resolution, especially at the lower frequencies. The floor dip just above 100Hz has been caught, but the RTA has averaged through the dip at 60Hz, which is narrower than 1/3-octave.

Figure 4.32 is a plot of the 1/3-octave GP response of the 180mm driver. Because the GP geometry eliminates the deep-floor reflections, the RTA does a good job of reproducing the 180mm driver GP low-frequency response. In particular, it shows the dips at 120 and 32Hz, which happen to line up well with 1/3-octave band-center frequencies. In both cases, the frequency response above 200Hz agrees with the plots of *Fig. 4.25*.

Comparing the staircase plots of *Figs. 4.31* and *4.32* with earlier continuous plots is somewhat difficult. If the measured 1/3-octave levels are plot-

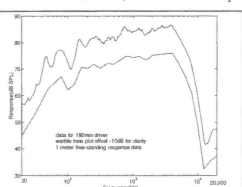

FIGURE 4.36: Comparison of warble-tone and smoothed sine-wave responses.

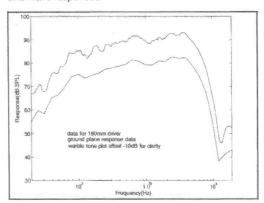

FIGURE 4.37: Comparison of GP warble-tone and smoothed sine-wave responses.

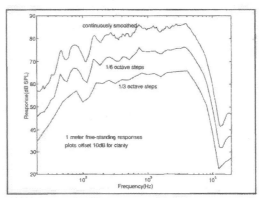

FIGURE 4.38: Smoothed sine-wave versus warble tones with 1/3 and 1/6 octave steps.

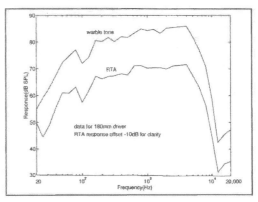

FIGURE 4.39: Comparison of warble tone and RTA one meter responses.

67

ted as *points* on a frequency-response graph and connected with straight lines, we get an easier plot to evaluate. *Figure 4.33* compares the one-meter RTA response plotted in this manner against the one-meter, 1/3-octave smoothed sine-wave response of our example 180mm driver. Now we see that the RTA has captured the general response shape, albeit with some of the finer detail missing. *Figure 4.34* shows the same comparison for the GP measurements.

We have examined its use in loudspeaker response measurement, but, as its name implies, the "real-time" analyzer is really best suited for

gular waveform, the instantaneous frequency will be linearly swept up and down around the unmodulated center frequency determined by the oscillator frequency control. The modulating frequency is typically set in the range of 4–5Hz. Played over a loudspeaker, this signal has a "warbling" quality, thus its name.

Adjusting the triangle wave amplitude so that the swept signal varies the frequency by $\pm 1/6$ octave gives us a 1/3-octave band signal with many advantages relative to an RTA. First, the signal is automatically limited to 1/3-octave (except for some very low-level FM sidebands), eliminating the need for a complex bank of 1/3-octave band filters. Because the signal has constant amplitude, an average responding AC voltmeter may be used and an averaging of multiple samples is unnecessary.

Warble-tone signal sources are considerably less expensive than RTAs. At the low end of the cost scale, CDs with 1/3-octave warble tones are available for about $20 US. A warbler oscillator kit is available from OCSL for about $120 US. This oscillator provides switched frequency outputs at 32 1/3-octave band center frequencies ranging from 16Hz–20kHz. Function generators with sweep-frequency capability are also widely available for about $200 US.

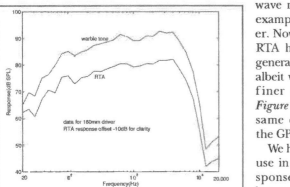

FIGURE 4.40: Comparison of warble tone and RTA ground-plane responses.

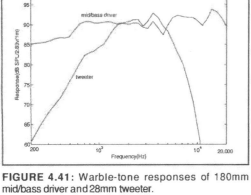

FIGURE 4.41: Warble-tone responses of 180mm mid/bass driver and 28mm tweeter.

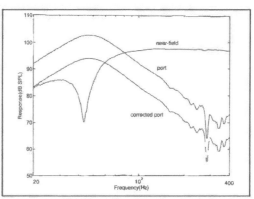

FIGURE 4.43: Near-field and port responses of 180mm mid/bass driver.

These oscillators have the added advantage of continuously variable center frequencies, enabling you to plot more than 31 points on the response curve for a more detailed result. The setup for warble-tone testing is shown in *Fig. 4.35*. Except for replacing the sine-wave oscillator with a warble-tone oscillator, this setup is no more complex than the sine-wave test setup shown in *Fig. 4.1*.

Figures 4.36 and *4.37* compare 1/3-octave warble-tone and smoothed sine-wave responses of our example 180mm driver in one-meter free-standing and ground-plane measurement scenar-

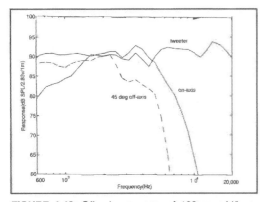

FIGURE 4.42: Off-axis response of 180mm mid/bass driver and 28mm tweeter.

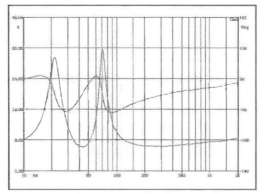

FIGURE 4.44: 180mm driver impedance plot.

jobs such as room/loudspeaker system response equalization. Using an RTA, together with a 1/3-octave band equalizer, it is possible to equalize a loudspeaker/room combination for flat frequency response in a matter of minutes. While not recommending against the use of RTAs, there is a simpler analog smoothing technique well-suited to loudspeaker response measurement using test gear that is relatively easy to build or less expensive to purchase.

4.5.2 THE WARBLER

If an audio oscillator is FM modulated with a trian-

ios. The results are similar to those obtained with the RTA. In particular, floor- and standing-wave-induced dips below 100Hz are missed in the one-meter warble-tone analysis.

The OCSL warbler was modified to produce 1/3-octave warble tones in 1/6-octave steps, doubling the number of measurement points. *Figure 4.38* compares one-meter 1/3-octave warble-tone responses taken in 1/3- and 1/6-octave steps against the smoothed sine-wave response. We see that doubling the number of 1/3-octave measurement points has allowed us to effectively recover all the significant response wiggles of the DUT.

Finally, 31-point warble-tone and RTA responses are compared in *Figs. 4.39* and *4.40*. Although the responses are quite similar, there are subtle differences due to the weight each approach gives to the data within each 1/3-octave band. The warbler equally weights all frequencies within a band, while the RTA gives more weight to the center frequency because of the shape of the bandpass-filter response curve.

4.6 TWO EXAMPLES

This section contains two interesting examples. The first illustrates how the frequency-response curves can be used to select a crossover frequency. The second describes an alternate and usually more accurate technique to determine a vented-box tuning frequency.

4.6.1 SELECTING CROSSOVER FREQUENCIES

This example shows you how to use warble-tone responses to select crossover frequency. *Figure 4.41* illustrates the warble-tone frequency response of the 180mm driver we have used in a number of previous examples, along with the response of a 28mm soft-dome tweeter mounted on the same baffle. The tweeter/woofer combination is free-standing with the tweeter at a height of 40″. The woofer axis is 7″ below the tweeter.

The microphone alternates from the tweeter to the woofer axes at a distance of one meter from the front baffle for each measurement. For these measurements the 1/3-octave warble tones are stepped in 1/6-octave increments for finer detail. The drive level to each transducer is maintained at 2.83V to guarantee that the relative sensitivities of the drivers are accurately captured.

Referring to the figure and, for the moment, ignoring the spreading loss below 600Hz, we see that woofer and tweeter responses are comparable in sensitivity and overlap in the 1.6–5kHz range. Based upon the on-axis data alone, you might choose a crossover frequency in the middle of the overlap region. In particular, a crossover somewhere in the 2.8–3kHz range might seem appropriate.

Figure 4.43, however, gives us a different picture. Here the curves of *Fig. 4.42* are reproduced along with the 45° off-axis response of the woofer in the horizontal plane. The low end of the frequency scale in this figure has been moved up to 600Hz to provide a little more detail. The 45° off-axis woofer response is falling rapidly above 2kHz. Although not shown, the tweeter horizontal coverage at 2kHz is almost ±90°.

Using a high-slope crossover at 3kHz with this driver pair will lead to suck-out in the off-axis response as the system transitions from the narrow polar response of the woofer, to the much broader horizontal coverage of the tweeter. This in turn will cause an undesirable mismatch in timbre between the direct- and reverberant-field responses. Moving the crossover frequency down to 2kHz will do much to even out the system off-axis response. Fortunately, the tweeter used in this example will handle the lower crossover frequency with ease.

4.6.2 NEAR-FIELD RESPONSES OF A VENTED SYSTEM

In this example, our 180mm woofer is placed in a 15 ltr vented box with a 10″ × 14″ front baffle. Very high resolution, near-field, sine-wave woofer and port responses are shown in *Fig. 4.43*. Notice the very deep notch in woofer acoustic output at 43.6Hz. This is a very accurate measure of the vented system tuning frequency. The tuned box/port combination places its maximum load on the woofer diaphragm at this frequency. Also notice that the port output peaks much more broadly, slightly above the woofer notch. In general, the peak port output frequency does not correspond to the system tuning and should not be used as a measure of this frequency.

As a double check on tuning, a high-resolution impedance plot for the system is shown in *Fig. 4.44*. The impedance minimum between f_L and f_H occurs at 44.2Hz—close, but not right on. The zero phase point is at 43.6Hz, which is also the acoustic notch frequency. The excellent zero-phase agreement is a direct consequence of the copper shorting disk used on the pole piece of this driver, greatly reducing the effect of voice-coil inductance. In general, impedance data and acoustic data do not agree that well. If this is the case, use the woofer notch frequency for the most accurate measure of system tuning.

Port peak acoustic output in the near-field is about 5dB higher than woofer output above 200Hz. This is an artifact of the near-field technique and the different port and cone areas. To get the correct half-space response of the port relative to the woofer we must correct for the different areas. The correction term is a function of the ratio of the port to cone diameters. The port diameter is 2″, while the effective diaphragm diameter of the 180mm driver is 5.45″. The correction term is:

$$\text{Correction} = 20\log_{10}\left(\frac{2}{5.45}\right) = -8.7\text{dB}$$

The corrected port response is also plotted on *Fig. 4.43*. Corrected port output in the notch region is over 20dB higher than the woofer output, indicating excellent reflex action. Unfortunately, we cannot add the port and woofer curves at other points to get the complete system response because we do not know the relative phase of the two outputs. For example, the port and woofer outputs are of equal magnitude at 20Hz, but because they are out-of-phase, the total output is actually very low.

4.7 FINAL COMMENTS

You have probably noticed the absence of any simple techniques for measuring driver phase response in this chapter. Analog techniques for measuring phase response are either expensive and tedious to apply or unreliable. One of the biggest problems with measuring phase response is locating a driver's acoustic phase center. This position is needed in order to remove the phase delay caused by the signal "fly time" from the driver to the microphone. Most dynamic drivers are minimum phase transducers at least over their useful frequency range. I will discuss the minimum phase property in more detail in Chapter 6, but for now, it is enough to know that for minimum phase devices there is a direct mathematical relationship between frequency response and phase response, called the *Hilbert transform*.

Relatively inexpensive software packages such as XOPT and SoundEasy can calculate driver phase response from your measured frequency response data using the Hilbert transform. Other programs, including Calsod and LMP, first model driver response with an RLC circuit which is then used to calculate phase response. Either approach will be more accurate and much faster than measuring phase response directly with simple instrumentation.

REFERENCES

1. J.A. D'Appolito and R.H. Campbell, "Mitey Mike II," *Speaker Builder*, 4/97. Available from Old Colony Sound Lab, PO Box 576, Peterborough, NH 03458, (603) 924-6371, FAX (603) 924-9464, E-mail custserv@audioXpress.com.
2. Beranek, *Acoustical Measurements*, Acoustical Society of America, revised edition, 1988.
3. J.R. Wright, "Fundamentals of Diffraction," *JAES*, Vol. 45, pp. 347–356 (May 1997).
4. J. Vanderkooy, "A Simple Theory of Cabinet Edge Diffraction," *JAES*, Vol. 39, pp. 923–933 (December 1991).
5. D.B. Keele, Jr., "Low-Frequency Loudspeaker Assessment by Nearfield Sound Pressure Measurement," *JAES*, Vol. 22, pp. 154–162 (April 1974).
6. C.J. Struck and S.F. Temme, "Simulated Free Field Measurements," *JAES*, Vol. 42, pp. 467–482 (June 1994).
7. M.R. Gander, "Ground-Plane Acoustic Measurement of Loudspeaker Systems," *JAES*, Vol. 30, pp. 723–731 (October 1982).

APPENDIX A
MEASUREMENT MICROPHONES[1]

In order for any instrument to perform accurate frequency response measurements, the first priority is a microphone with a reliably flat frequency response. The following is a list, descending in price, which are acceptable as measurement microphones. The first four are precision measurement microphones, mostly used by professionals, and priced between $3,500 and $1,600 for microphone capsule, preamp body, and power supply. The last four are low-cost microphones which have sufficient precision to be included in this grouping and range in price from $550–$230.

PRECISION MICROPHONE LISTING

Bruel & Kjaer Model 4133
Rion Industries Model UC-31
Larson-Davis Model 2540
ACO Pacific Model 7012

AKG C460B (CK 62 capsule)
Josephson Engineering
Neutrik 3382
Mitey Mike II (Old Colony Sound)

In the groups, two microphones stand out as particularly good "buys." In the first group of professional precision microphones, the ACO Pacific represents an excellent value. It performs well in comparison to other precision microphones and is priced well under the competition. Among the group of less expensive microphones, the Mitey Mike II designed by Joe D'Appolito and sold by Old Colony Sound Lab is as good as it gets for the price.

[1](From *Loudspeaker Design Cookbook*, p. 139).

CHAPTER 5

ACOUSTICAL TESTING OF MULTIPLE DRIVER SYSTEMS

5.0 INTRODUCTION

Perhaps the single most important measure of loudspeaker-system quality is frequency response, both on-axis and off-axis. In this chapter we will extend our discussion of acoustical testing to encompass frequency-response measurements of multiple-driver systems. All of the environmental considerations covered in Chapter 4 in connection with single-driver tests apply to the testing of multiple driver systems. There are at least two additional issues unique to multiple-driver systems—driver integration and multiple-driver floor bounce.

After discussing these issues I will present several examples of frequency-response measurements on two- and three-way systems using dynamic drivers and a very interesting two-way system using a wideband ribbon driver. In addition to on-axis measurements, I will show the polar response of several of the examples and comment on the impact this polar response may have on perceived spectral balance in typical listening rooms. You will also see the response of the individual drivers in a system context and the effect of the crossover networks. The chapter will end with a brief discussion of power response illustrated with a subwoofer example.

5.1 MULTIPLE-DRIVER INTEGRATION

A representative driver placement for a three-way loudspeaker system is shown in *Fig. 5.1*. The tweeter is typically placed at ear height for a seated listener. The midrange unit is usually placed just below the tweeter, with the woofer below the midrange and often close to the floor. Because the acoustic centers of the individual drivers are not coincident, the distances from each driver to the listening or measurement position are not the same.

For the microphone location shown in *Fig. 5.1*, the midrange and woofer drivers will lag behind the tweeter both in time and phase. Designers often pick a listening location and then design the crossover network, with a possible tilt of the enclosure or offset of driver positions, to optimize the frequency response at the chosen location. For the home listening room a seated ear height of 36″–38″ at a distance of 3–4m represents a good design point.

From *Fig. 5.1*, we see that testing multidriver systems poses a problem. If we place the microphone too close to the system, the drivers may not integrate properly. If we place the microphone too far from the system, our measurements of the direct field may be highly contaminated with room reflections. To assess the magnitude of this problem, let's look at a specific example.

Our example will examine the effect of microphone location on measured system response using both even- and odd-order crossover networks. Odd-order networks, such as first- and third-order net-

works, produce a 90° phase shift between drivers at the crossover frequency. With even-order networks like the second- and fourth-order Linkwitz-Riley crossovers, the inter-driver phase shift is zero; that is, the drivers are in phase at crossover.

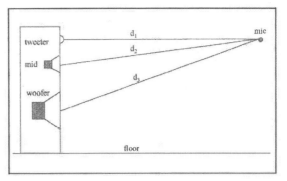

FIGURE 5.1: Driver integration geometry.

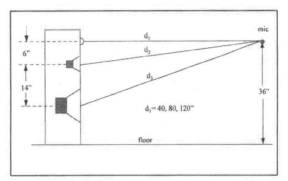

FIGURE 5.2: Example of driver integration geometry.

TABLE 5.1				
EFFECT OF MICROPHONE LOCATION ON WOOFER/MIDRANGE INTEGRATION				
Microphone distance (in)	Path-length difference (in)	Added phase shift (degrees)	Response dip (dB) odd-order (at 500Hz)	even-order
40	4.3	58	8.1	1.2
80	2.2	30	3.0	0.3
120	1.5	20	1.8	0.1

TABLE 5.2				
EFFECT OF MICROPHONE LOCATION ON MIDRANGE/ TWEETER INTEGRATION				
Microphone distance (in)	Path-length difference (in)	Added phase shift (degrees)	Response dip (dB) odd-order (at 3000Hz)	even-order
40	0.45	36	3.9	0.44
80	0.22	18	1.6	0.11
120	0.15	12	1.0	0.05

Figure 5.2 is a copy of *Fig. 5.1* with specific dimensions added. The tweeter-to-midrange spacing is 6″. Midrange-to-woofer spacing is 14″. The microphone is placed at tweeter height at distances of 40″, 80″, and 120″ from the tweeter. In practice, the midrange and woofer acoustic centers are behind the tweeter when all of the drivers are mounted flush on a common baffle.

In Chapter 7 we will learn how to measure driver acoustic-center location. For this example, however,

I will ignore the acoustic-center offsets and assume all drivers are aligned vertically at the face of the enclosure. Including the acoustic-center offsets would make the results of this example worse. Tweeter/midrange and midrange/woofer crossover frequencies are 3000Hz and 500Hz, respectively.

Table 5.1 lists the path-length difference to the microphone, the additional phase shift caused by this difference, and the dip in measured direct-field response caused by this phase shift at the woofer/midrange crossover frequency as a function of microphone location. Floor and wall reflections are not considered in this example. *Table 5.2* is a similar listing for the midrange/tweeter pair.

Two conclusions are evident from these tables. First, as the microphone distance increases, the path-length difference decreases with a corresponding improvement in driver integration. This is simple geometry. Second, at any distance, odd-order crossovers cause at least an order of magnitude more response error than even-order networks.

At 40″ (approximately 1m) the response dips of 8.1 and 3.9dB caused by odd-order networks are clearly unacceptable. The even-order dips of 1.2 and 0.44dB could easily be missed or confused with other response irregularities caused by surface reflections or the drivers themselves. This is one reason why many designers prefer even-order crossovers. They are much less sensitive to listener location. A more general conclusion from these tables is that measured multiway loudspeaker frequency response is critically dependent on microphone location. We'll see many more examples of this dependence in later sections of this chapter.

Some authors[1] cite a distance of three times the largest dimension of the loudspeaker as a minimum distance for good driver integration. They refer to this distance as the "far field" for the system. Based on this example, a distance of more like six times the driver span is needed for odd-order crossovers.

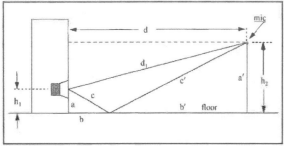

FIGURE 5.3: General floor-bounce geometry.

5.2 FLOOR BOUNCE REVISITED

In Chapter 4 we examined the floor bounce for the simple case where driver and microphone were both at the same height off the floor. With multidriver systems, the driver and microphone will generally be at different heights. A more general case for floor-bounce geometry is shown in *Fig. 5.3*. The driver is at height h_1 and the microphone

is at height h_2. The microphone and baffle are separated by a distance, d. The direct path is labeled d_1. The bounce path is made up of two segments, c and c′.

Our problem is to determine the path-length difference $(c + c') - d_1$. Once we know this difference we can use the equations of Chapter 4 to compute the response dips and peaks due to floor bounce. To do this we need a little geometry.

Let's determine c and c′ first; h_1, h_2, and d are known. Triangles abc and a′b′c′ are *right* triangles. Furthermore, from ray theory we know that the angle of incidence equals the angle of reflection. That is:

$$\angle bc = \angle b'c'$$

Since the two triangles have two equal angles, the third angles must also be equal, and triangles abc and a′b′c′ are similar. That means:

$$\frac{b'}{b} = \frac{a'}{a} = \frac{h_2}{h_1} = v \qquad [5.1]$$

and

$$b' = d - b \qquad [5.2]$$

Combining Equations 5.1 and 5.2 gives us

$$b' = \frac{v \times d}{v + 1} \qquad [5.3]$$

Once we have b′, we can compute c′ and b. Then, knowing b we can compute c. The appropriate equations are listed below:

$$c' = \sqrt{h_2{}^2 + \left(b'\right)^2} \qquad [5.4]$$

$$b = d - b' \qquad [5.5]$$

and

$$c' = \sqrt{h_1{}^2 + \left(b\right)^2} \qquad [5.6]$$

The equations for c and c′ follow from the fact that both triangles are right triangles. Finally, notice that the dashed horizontal line in *Fig. 5.3* together with the baffle face and the line segment d_1 form another right triangle. This allows us to compute d_1 as follows:

$$d_1 = \sqrt{d^2 + \left(h_2 - h_1\right)^2} \qquad [5.7]$$

Let's apply these equations to the midrange and woofer drivers of *Fig. 5.1* using a microphone spacing of 80″. For the woofer, the dimensions are:

$$d = 80''$$
$$h_1 = 16''$$
$$h_2 = 36''$$

Then:

$$v = \frac{36}{16} = 2.25$$

$$b' = \frac{2.25 \times 80}{2.25 + 1} = 55.38''$$

$$c' = \sqrt{36^2 + 55.38^2} = 66.06''$$

$$b = 80 - 55.38 = 24.62''$$

$$c = \sqrt{16^2 + 24.62^2} = 29.36''$$

$$d_1 = \sqrt{(36-16)^2 + 80^2} = 82.46''$$

and finally
$$(c + c') - d_1 = 29.36 + 66.06 - 82.46 = 12.96'' = 0.329m$$

Using Equation 4.3 of Chapter 4, we find that the frequency of the first floor-induced response dip is at

$$f = \frac{180}{0.344} = 522.8 Hz$$

Repeating these calculations for the midrange driver placement gives a path-length difference of 0.596m and a first-floor bounce dip at 288.6Hz. Notice that the first midrange floor dip is almost an octave below the first woofer dip. We can use this spread in dip frequencies to great advantage.

> **DESIGN TIP!** Placing the woofer/midrange crossover frequency between the two first dips, preferably at their geometric mean, will greatly reduce or eliminate these dips from the system frequency response. That is:
>
> crossover frequency = $\sqrt{522.8 \times 288.6} = 388.4 Hz$
>
> With this crossover frequency the woofer will be rolling off before it reaches its first-floor dip. Likewise, the midrange output will be well down in response at its first-floor dip frequency.

Before leaving this section, I should point out that ground-plane measurements have many advantages in measuring multiway systems. First of all, the inherent 6dB signal gain of this technique allows you to place the mike at greater distances from the system under test (SUT). This greatly mitigates the driver integration problem discussed in Section 5.1. Also, the floor bounce is eliminated and wall reflections are generally reduced.

5.3 TWO-WAY SYSTEM EXAMPLES

In this section we will examine the frequency response and impedance of three two-way systems. The test setup in all cases is similar to that shown in *Fig. 4.35*. The frequency source for all *acoustic* measurements in this chapter is a sine-wave oscillator frequency modulated to produce 1/3-octave-wide warble tones. The oscillator used in these tests is a

modified version of the Old Colony Sound Lab Warbler.[2]

The first modification enables the 1/3-octave warble tones to move in 1/6-octave steps to provide smoother response plots. The second change adds a switch to disable the warble-tone modulator to give a pure tone and replaces the stepped frequency control with a continuously variable fine frequency control. This change is useful for near-field frequency response, impedance measurements, and crossover testing. A calibrated laboratory microphone is used in all of the tests to be described, together with a wideband preamp and a lab-grade AC voltmeter.

5.3.1 EXAMPLE 5.1: A CLOSED-BOX TWO-WAY SATELLITE SYSTEM

In this example I will first describe the individual driver data taken to design the system. This data is used to select an appropriate crossover frequency and to design the crossover network for each driver. Then we will examine the measured response of the resulting system.

The driver complement for this example consists of a 170mm cellulose pulp cone woofer and a 3″ ribbon tweeter mounted in an 18 ltr closed box. The front baffle dimensions are 15″h × 9″w. The drivers are mounted on the baffle centerline, 6.5″ apart center-to-center. The box is fully stuffed with Dacron™ fiber.

As part of the design process, the impedance of

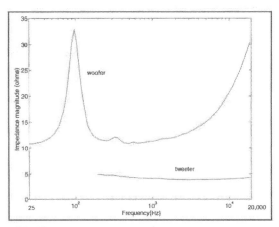

FIGURE 5.4: Driver impedance plots of Example 5.1.

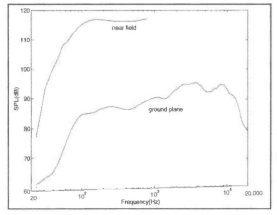

FIGURE 5.5: Woofer 2m ground-plane and near-field responses of Example 5.1.

both drivers was measured using the voltage divider procedure of Chapter 2. A sufficient number of points were taken with a variable frequency sine-wave oscillator to produce smooth impedance curves. The data was then plotted with an engineering software package.

The results are shown in *Fig. 5.4*. The woofer has a closed-box resonant frequency of 97Hz with a Q of 0.9. The tweeter impedance is an almost constant and nearly resistive 3.9Ω.

Frequency response of the woofer when mounted in the enclosure is shown in *Fig. 5.5*. Both 2m ground-plane and near-field responses are shown. Recall from Chapter 4 that the 2m ground-plane

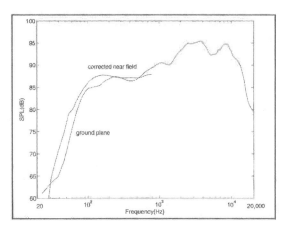

FIGURE 5.6: *Figure 5.5* with corrected near-field response.

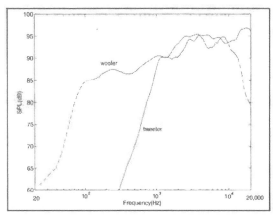

FIGURE 5.7: Woofer and tweeter responses of Example 5.1.

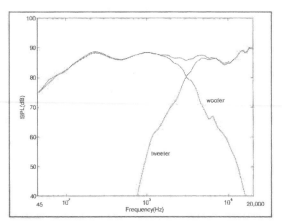

FIGURE 5.8: Example 5.1's 2m ground-plane system and driver responses.

measurement produces the same sensitivity as a 1m free-standing response. The nearest reflecting surfaces were at least 20′ away, guaranteeing that all reflections are down at least 16dB.

Although the woofer's nominal sensitivity rating is 93dB, we see that diffraction spreading loss has reduced the on-axis sensitivity to about 87dB below 500Hz. For a flat on-axis response, this lower figure sets the overall system sensitivity.

Turning to the near-field response, the effective diameter of the woofer is 14cm. The upper frequency limit of the near-field measurement computed using Equation 4.4.1b is:

$$f_{MAX} = \frac{10,950}{14} = 782 Hz$$

Since the woofer is radiating into a 4π space below 500Hz, we should correct the near-field measurement to 1m using Equation 4.4.2a. Using 7cm or 0.07m for the driver radius, the value of the correction term is:

$$20\log_{10}\left(\frac{4 \times 1}{.07}\right) = 35.1dB$$

This would be the right value for the correction term if the drive level in both tests had been the same. In order to avoid microphone preamp overload, however, I reduced the drive level during near-field testing by 10dB. Thus we must correct the correction term by 10dB to 25.1dB. Subtracting 25.1dB from the near-field data produces the response plot shown in *Fig. 5.6*. The 2m ground-plane and near-field curves are in good agreement. Because the ground-plane data is relatively free of reflections, it does a pretty good job of characterizing low-frequency response.

Woofer and tweeter responses are compared in *Fig. 5.7*. Tweeter sensitivity on this baffle averages 94dB between 1 and 3kHz and rises to about 97dB above 15kHz. Below 1kHz tweeter response falls with decreasing frequency at 12dB/octave. Woofer and tweeter responses overlap between 1 and 10kHz. This overlap suggests that a crossover frequency around 3kHz would work well. This also jibes well with the manufacturer's recommendation of a 3kHz crossover frequency for the tweeter with at least a third-order electrical crossover.

One important comment on the woofer/tweeter pair measurement: In addition to the frequency response of each driver, this measurement also gives us driver sensitivity. If you have a calibrated microphone and a known constant drive voltage, you can determine the absolute sensitivity of each driver. If your microphone is not calibrated, but known to be reasonably flat, you can still get the relative sensitivity of each driver. This is best accomplished by keeping the drive voltage to the woofer and tweeter the same during these tests. Then no additional correction is needed for the different drive levels.

This book is not about crossover design, so I will not cover the details of the crossover design for this system here. However, because of the intimate relationship between driver measurements and crossover design, I will summarize the approach I used. Today there are available a number of crossover optimization programs which greatly enhance the crossover design process, both in terms of speed and accuracy. Examples of such optimization programs are XOPT™, LEAP™, and SoundEasy™.

The measured frequency response and impedance magnitude data for both drivers are input to a crossover optimization program. In addition to the data already collected, both driver phase response and the impedance phase angle are needed to perform the optimization. Without phase data the interaction between the crossover network and the driver voice-coil impedance cannot be computed. Nor can the driver acoustic responses be added properly.

We do not have phase data. Fortunately, the programs cited above compute driver phase response from frequency-response data using a mathematical operation called the Hilbert transform. This operation assumes the drivers are minimum phase. I will define the minimum phase property in Chapter 6 and examine this assumption relative to drivers in Chapter 7. The programs also fit an RLC model to the impedance magnitude. Impedance phase is then derived from this model.

To start the optimization process you must input a target sensitivity, a target acoustic response function, a crossover topology, and starting values for the crossover components. You must also specify

the design axis or axes and driver locations relative to the axis or axes. The software will then optimize the component values to move the system response as close to the target response as possible within the constraints of the chosen circuit topology.

This all sounds pretty straightforward. Be aware, however, that crossover optimization software is rather dumb. None of the software that I know of will accept constraints on system impedance or limits on component values. A great deal of operator intervention is required to "steer" the optimizer and obtain a useful design.

Getting back to our first example, the overall sensitivity of this system is determined by the woofer response below 500Hz. This sets the target sensitivity at 87dB/2.83V/1m. A fourth-order Linkwitz-Riley *acoustic* crossover at 3kHz was chosen as the target function for each driver.

This means that the drivers should be in phase at the crossover frequency with each driver's acoustic response down 6dB relative to the system response. Furthermore, the acoustic response of the woofer will fall at 24dB/octave well above the crossover frequency. Likewise, the tweeter's acoustic response will rise at 24dB/octave well below crossover.

I have discussed this point before, but it bears repeating. In the previous paragraph, the word *acoustic* in connection with the crossover appears several times. The overall acoustic response will almost invariably be attained as a combination of the electrical crossover characteristics working in

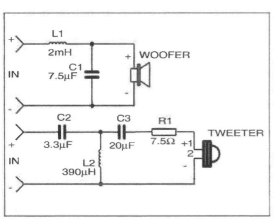

FIGURE 5.9: Crossover network for Example 5.1.

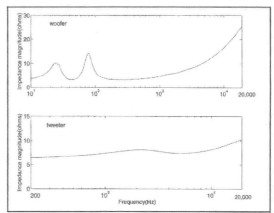

FIGURE 5.11: Example 5.1: Crossover voltage responses.

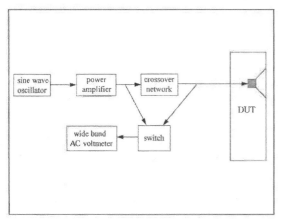

FIGURE 5.10: Test setup for crossover voltage response.

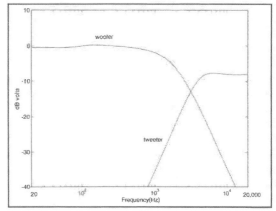

FIGURE 5.12: Example 5.2's woofer and tweeter impedance, respectively.

75

concert with the driver's natural acoustic response.

Figure 5.8 shows the ground-plane response of the final system. The microphone was placed on the floor along the tweeter axis at a distance of 2m. This was the design axis for the crossover optimization. The measured crossover frequency is 3050Hz. Response is flat within ±2dB between 100Hz and 10kHz.

Relative to the system response, both drivers are down 5.7dB at crossover. This translates to an interdriver phase angle of 14°. This is very close to the ideal in-phase crossover and, practically speaking, this is about as good as you can get. A 14° phase shift at 3050Hz corresponds to a path-length difference of only 0.17″, or 0.43cm. Moving the micro-

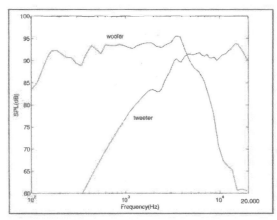

FIGURE 5.13: Free-standing woofer and tweeter responses of Example 5.2.

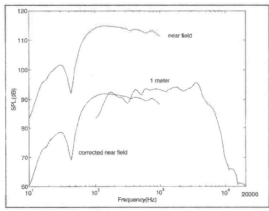

FIGURE 5.14: Woofer responses of Example 5.2.

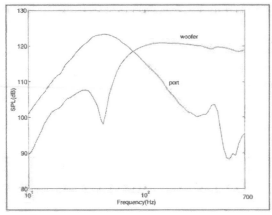

FIGURE 5.15a: Example 5.2's woofer and port near-field responses.

phone below the tweeter axis by about 13cm will put the drivers in phase at 2m. Again, we see how important microphone placement is.

The crossover network for this example is shown in *Fig. 5.9*. Notice that the woofer and tweeter crossovers are second- and third-order electrical filters, respectively, and yet the system achieves an overall fourth-order *acoustic* response. This is one of the great advantages realized in using optimization software. The software takes full account of the electrical impedance and acoustical response of the drivers and incorporates these characteristics into the design.

Take the woofer as an example. *Fig. 5.4* shows the woofer impedance rising rapidly above 1kHz. From earlier discussions we know that this impedance looks like a lossy frequency-dependent inductor. Conventional wisdom says that the voice-coil impedance rise should be compensated with a Zobel network to make it look purely resistive.

This would then allow us to design a "textbook" crossover for a driver with a constant purely resistive impedance. This crossover would be fourth-order with two inductors and two capacitors plus the Zobel. However, by incorporating the effect of the lossy inductor into the crossover design and also accounting for driver frequency response and phase shift, we can get the fourth-order acoustic crossover with a second-order electrical filter! We have saved four components, two from the crossover and two from the Zobel.

The crossover voltage response is also of interest. *Figure 5.10* shows the test setup for this measurement. *The crossover network and drivers are connected to the amplifier at all times during the test.*

Assuming the sine-wave oscillator and amplifier responses are flat over the frequency region of interest, first place the AC voltmeter test leads across the amplifier output and adjust the oscillator output to give a reading of 0dB on the 0.3V or 1V scale. (Some AC voltmeters have one input terminal grounded. Be careful not to short the output of your amplifier to ground. It may go up in smoke!) Use the lowest level consistent with reliable measurements, since the sine-wave tones coming from the DUT can become quite annoying.

Once the reference level is set, move the voltmeter test leads to the crossover output terminals and measure the frequency response of the crossover using the decibel scale. Occasionally check the amplifier output to make sure the 0dB reference has not changed. For woofers, vary the frequency from 20Hz up to a point at least two octaves above the expected crossover frequency. Take enough points to plot a smooth curve. For tweeters, begin your test at least two octaves below the expected crossover frequency and continue on up to 20kHz.

The voltage responses of the woofer and tweeter networks are shown in *Fig. 5.11*. The woofer network response starts rolling off above 500Hz and is down about 11dB at 3kHz. Contrast this with a textbook fourth-order crossover response which would be down only 6dB at 3kHz. The early rolloff is required to counteract the rising response of the woofer above 500Hz.

The tweeter network produces an attenuation of 7dB in its passband. This is needed to reduce the 94dB tweeter level down to the overall system sensitivity of 87dB. Because the tweeter impedance is essentially a constant resistance of 3.9Ω, this attenuation can be accomplished with a single series resistor.

5.3.2 EXAMPLE 5.2: A TWO-WAY, TWO-DRIVER VENTED SYSTEM

5.3.2.1 ON-AXIS RESPONSE

This example couples a 180mm midbass driver with a 25mm textile dome tweeter. The woofer has a Neoprene surround and a cellulose cone coated with a butyl damping compound. The enclosure is a vented box of 15 ltr internal volume lined with egg-crate acoustic foam on three sides and the top and bottom. A 2.25″ ID vent tunes the enclosure to 43Hz. Impedance plots for both the woofer and tweeter are shown in *Fig. 5.12*. The tweeter resonant frequency is approximately 2kHz. Heavy ferrofluid damping of the tweeter voice coil makes determining the exact frequency somewhat difficult.

Freestanding 1m frequency responses for both the woofer and tweeter are plotted in *Fig. 5.13*. The tweeter is rolling off on the low end below 3kHz at 12dB/octave with decreasing frequency. Woofer response peaks at a level of 95dB at 4kHz and falls off above that frequency at 12dB/octave also. There is very little frequency overlap.

Floor reflections corrupt the woofer response

curve below 450Hz, making it difficult to evaluate the spreading loss. A ground-plane measurement could clear this up, but *Fig. 5.14* illustrates another approach. Here we see the woofer 1m and near-field plots on the same graph. Correcting the near-field response to 1m will give us a good estimate of the sensitivity below 500Hz. The woofer in this example has the same effective cone radius as the one in Example 5.1.

As in the first example, the total 4π near-field to 1m correction is 35.1dB, but the microphone preamp gain was reduced again by 10dB for the near-field measurement, so we must subtract only 25.1dB from the near-field data. The corrected near-field curve is also plotted on *Fig. 5.14*. From this curve we see that the low-frequency sensitivity of the woofer is roughly 90dB.

The near-field data in this example was taken with a continuously variable frequency sine-wave oscillator for higher resolution. As discussed in Section 4.4.2, smoothing of near-field measurements is not required because of the very high signal-to-noise ratio. The sharp dip in the near-field response plot clearly shows the port tuning frequency. The near-field plots generally give us a much better measure of f_B than we can get from the impedance curve.

The near-field woofer response is not a good indicator of low-frequency extension, however, because most of the output is coming from the port at this point. Near-field port and woofer responses are plotted in *Fig. 5.15a*. Both curves were made at

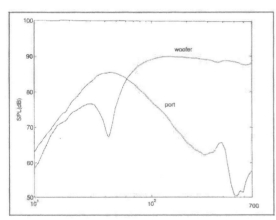

FIGURE 5.15b: Near-field responses corrected to 1m.

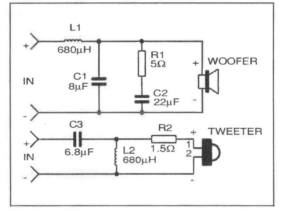

FIGURE 5.17: Crossover network for Example 5.2.

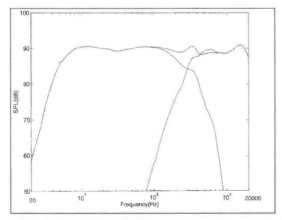

FIGURE 5.16: Example 5.2's system and driver responses.

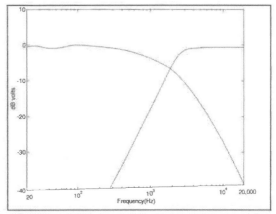

FIGURE 5.18: Crossover voltage responses of Example 5.2.

the same woofer drive level. At its peak, the port response appears to be 2–3dB higher than the woofer response above 100Hz. At first glance you might suspect that the system will have an undesirable peak in bass response around 45Hz.

However, the equations which correct these near-field responses to 1m involve the source radius. We already know the woofer response correction is 25.1dB. The port radius is 1.125″, or 0.0286m. Using this value in Equation 4.4.2a yields a correction term for the port output of 32.9dB (actually, 42.9dB less 10dB for the mike preamp gain reduction). The difference in correction terms is 7.8dB. Notice that this difference is simply the ratio of the port to woofer radii or diameters.

$$\text{difference} = 20\log_{10}\left(\frac{2.25}{5.51}\right) = -7.8\text{dB}$$

The corrected woofer and port responses are plotted in *Fig. 5.15b*. Two points can be made here. First, at f_B the port output is almost 20dB above the woofer output so that the port level is a good measure of total system response at f_B. Second, the corrected port response at f_B is about 5dB below the woofer response above 100Hz. This is typical of a QB3 vented system alignment where the –3dB point is generally above f_B.

Unfortunately, we cannot add the woofer and port outputs directly to get a more complete picture of low-frequency response. To add them we must know phase in addition to level. We will see how to do this with PC-based acoustic data acquisition hardware and software in Chapter 7.

A fourth-order acoustic Linkwitz crossover was chosen for this system. After crossover optimization the final system response is shown in *Fig. 5.16*. This is a measured 2m ground-plane response above 200Hz. Below 200Hz a QB3 response model was joined to the measured response curve at 200Hz to get a complete response curve. The QB3 model was derived from vented-box measurements described in Chapter 3. Overall system response is flat within ±2dB from 200Hz to 18kHz.

Crossover occurs at 2900Hz. A schematic of the crossover network is shown in *Fig. 5.17*. Recall that both the woofer and tweeter have 12dB/octave response slopes near the crossover frequency. The optimization software incorporated these slopes into the overall fourth-order response. That is why both crossovers are only second-order electrical filters. The effect of woofer voice-coil impedance was not needed for this crossover so a Zobel was added to cancel it out. *Figure 5.18* shows the crossover network voltage response.

The impedance plot for the finished system is shown in *Fig. 5.19*. This plot was made using the voltage-divider technique described in Chapter 2. In addition to the vented-system double impedance peaks below 100Hz, there is a third large peak at 2kHz. This peak is caused by the rising inductive impedance of the woofer crossover network and

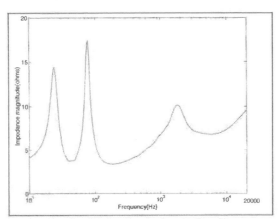

FIGURE 5.19: Example 5.2's system impedance.

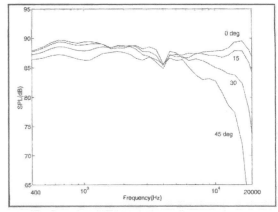

FIGURE 5.21: Horizontal polar response of Example 5.2.

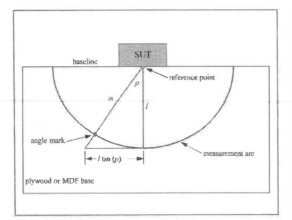

FIGURE 5.20: A polar-response test fixture.

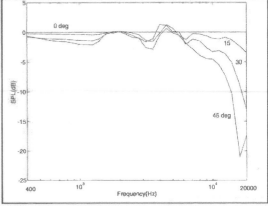

FIGURE 5.22: Example 5.2's normalized horizontal polar response.

the falling capacitive impedance of the tweeter crossover. These two impedance functions meet at 2kHz to form a parallel resonance.

5.3.2.2 POLAR RESPONSE

Before showing polar-response results for this example, let's discuss how polar-response data is obtained. The axis of rotation for horizontal polar response is usually chosen as the centerline of the baffle if the baffle is vertical. For tilted baffles, a vertical line through the tweeter diaphragm is a suitable rotation axis. You must be careful to keep the microphone-to-rotation-axis distance constant. Otherwise, changes in response due to microphone-to-driver movement may be confused with polar-response effects.

Invariably, there will be small changes in response because the system driver acoustic centers will not lie on the axis of rotation. In general, the horizontal polar response of tilted baffle systems will have significant distance effects mixed in with off-axis effects. This is the price paid for aligning driver acoustic phase centers.

In laboratory or production situations where a large number of polar response measurements are normally made on a daily basis, it is common practice to mount the SUT on a motor-driven turntable, often under computer control. One such turntable is available from Old Colony Sound Lab for about US $1700. The turntable is driven through fixed angular steps while the microphone position is held fixed. Steps as small as 5° are common. Frequency-response data is collected at each step. You will see polar-response data taken this way in Chapter 7.

You could construct a manually or electrically driven turntable. The turntable must be quite sturdy. Not only does it support the weight of the SUT, but it also must support a fairly large torque because the axis about which the SUT is turned rarely coincides with its center of gravity. Alternatively, you could simply place the SUT on the floor or on a suitable stand, mark off angular increments on the floor, and rotate the system by hand.

Figure 4.2 suggests a different approach. Here the SUT is held fixed and the microphone is moved. An obvious alternative, it does have some advantages in its implementation. *Figure 5.20* illustrates a simple scheme for making polar-response measurements this way.

Take a $4' \times 8'$ sheet of plywood or MDF and mark a reference point at the center of one long edge. Draw a measurement arc on the sheet centered on the reference point. The microphone will be positioned on or over this arc. A 1m radius is typical, although you could go out a bit farther. You can make a crude compass with a strip of wood a little over 1m long. Drive a nail through one end of the strip and into the reference point. At 1m away, drill a hole big enough to accept a broad-point marking pen. Insert the pen into the hole and rotate.

Use a large framing square to draw a line perpendicular to the reference point. This is line l in *Fig. 5.20*. This line defines the on-axis measurement location. You can then use a protractor to mark off angles along the measurement arc.

Alternatively, *Fig. 5.20* shows another way to determine the desired angles. If p is the angle and l is the length of l, mark off a distance $l \tan(p)$ from the end of l parallel to the baseline. Draw the line m. The angle you want will now be at the intersection of m with the measurement arc.

A big advantage of this simple fixture is that it can be used for both horizontal and vertical polar-response measurements. For horizontal polar-response measurements, place the SUT standing upright along the baseline as shown in *Fig. 5.20*. Put the microphone on a suitable stand bringing it to tweeter height. Then place the microphone stand sequentially on each angle mark and take a measurement.

Using the same fixture, vertical polar response is obtained with a ground-plane measurement. Lay the SUT on its side with the tweeter positioned at the reference mark. Lay the microphone on the fixture base and move it along the measurement arc taking response data as you go. Just be sure to tilt the speaker cabinet so that the tweeter axis intersects the measurement arc.

With these preliminaries over, let's examine the polar response of Example 5.2. Horizontal polar response for the example is shown in *Fig. 5.21*. The response on-axis (0°) is plotted for reference along with horizontal angles of 15°, 30°, and 45° right. For clarity, only four angular positions are shown. Otherwise, the plot quickly becomes cluttered.

Even with just four curves, the lines below 4000Hz are hard to separate visually. Because the drivers are aligned vertically on the baffle centerline, the corresponding angles to the left should yield the same response curves.

Below 5kHz, response is quite uniform, varying by no more than ±1.5dB. The tweeter becomes progressively more directional above this frequency and at angles greater than ±15°. Normally, a 25mm tweeter would show a broader polar response, but the diaphragm of this tweeter is recessed into its faceplate, shadowing its response at larger angles.

Figure 5.21 does not show polar response in its most useful form. In a good design, off-axis response should be a smooth replica of the on-axis response. Decreasing SPL levels with increasing angle and a progressive, though not excessive, rolloff of the highs are acceptable. Large variations in off-axis response relative to the on-axis curve are not.

The overall frequency balance of a loudspeaker as perceived by a human listener is a combination of direct and reflected sound. Off-axis energy arrives at the listening position after reflections off the walls. In typical listening rooms this energy arrives in a time span well within the Haas fusion zone,[3] a time interval starting just after first arrival and extending out to 40–50ms. Within this time interval the human hearing mechanism merges reflected and direct sounds. Even if the on-axis response is flat, poor off-axis response could produce unsatisfactory sound.

Thus we are really interested in the *change* in off-axis response relative to the on-axis response. If the on-axis curve is subtracted from each of the off-axis curves, we get the normalized horizontal polar response shown in *Fig. 5.22*. Here the on-axis plot appears as a straight line at 0dB. The remaining curves show the *differences* we are interested in. From this figure we more clearly see that the 15° line is within 1dB of the on-axis response at 10kHz. The 30° and 45° curves are down 2.5 and 5dB, respectively, at 10kHz.

In *Fig. 5.22*, the region around 3kHz is especially interesting. This is the crossover region. With a fourth-order crossover, the transition from the 180mm woofer to the 25mm tweeter is quite rapid. The woofer is becoming directional at this frequency, whereas the tweeter response

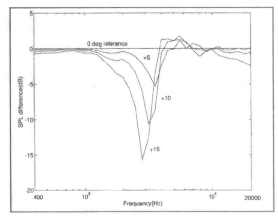

FIGURE 5.23a: Normalized vertical polar response above tweeter axis of Example 5.2.

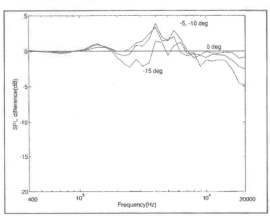

FIGURE 5.23b: Example 5.2's normalized vertical polar response below tweeter axis.

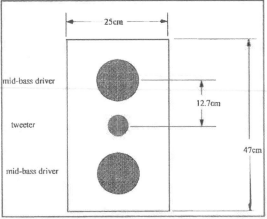

FIGURE 5.24: The MTM array.

is quite broad here. This is why we see a dip in off-axis response just below 3kHz, followed by a peak in off-axis response just above this frequency. The worst dip-to-peak span is only 4dB. This really isn't too bad.

Normalized vertical polar response is shown in *Fig. 5.23a* for angles of 5°, 10°, and 15° *above* the tweeter axis. All curves show a sharp dip around the 3kHz crossover frequency. The dips at 10° and 15° are probably deeper than shown. The 1/3-octave warble tones smooth through the true bottom of each dip.

A half wavelength at 3kHz is about 5.7cm. The woofer is below the tweeter by 16.5cm. Furthermore, its acoustic center is 2.5cm behind the tweeter. (This was determined using the techniques of Chapter 7.) Moving above the tweeter, the differential path length increases rapidly. At +11.1° the path length difference to the baffle surface is 3.2cm. Add the additional 2.5cm acoustic offset and you have a 180° phase shift.

Figure 5.23b shows the normalized vertical polar response for angles 5°, 10°, and 15° *below* the tweeter axis. The picture is much better here. Moving below the tweeter axis tends to bring the driver acoustic centers into alignment. The −15° curve shows the least difference with the on-axis response. Over the 15° range the worst-case difference is only 4dB. Clearly this system should be tilted back 8–10° for best vertical coverage. This puts the worst-case vertical response aiming up 20–25°. Fortunately, we are not too sensitive to arrivals from above.

5.3.3 EXAMPLE 5.3: TWO-WAY, THREE-DRIVER MTM SYSTEM[5,6,7]

5.3.3.1 ON-AXIS RESPONSE
Our next example involves a high-quality two-way monitor with two midbass drivers and a tweeter in an MTM array. For those readers not familiar with this configuration, an MTM array is shown in *Fig. 5.24*. A pair of woofers or midbass drivers are arranged symmetrically about a central tweeter. All drivers are aligned vertically. As we will see shortly, this layout and the proper crossover guarantee a smooth, symmetric vertical polar response.

In this example there are two 130mm midbass drivers. They use a polykevlar sandwich cone with a half-roll Neoprene surround mounted in a cast aluminum frame. The 28mm tweeter contains a concave crimped titanium diaphragm with a foam surround allowing exceptional excursion for a tweeter. Baffle dimensions and driver spacing for this example are shown in *Fig. 5.24*. The vertical baffle edges are heavily rounded. A 2″ ID port placed on the rear wall of the enclosure tunes the system to approximately 48Hz.

Woofer, port, and tweeter data were first collected for the system design. Woofer and port responses are shown in *Fig. 5.25a*. The freestanding response of the woofer pair above 300Hz was measured at 1m with the microphone placed along the tweeter axis. Near-field responses of a *single* woofer and the port were also taken to provide data below 300Hz. The woofer near-field response clearly shows the port tuning frequency.

Woofer and port near-field data were not taken at the same time as the woofer pair 1m data. Furthermore, the drive levels for both near-field responses were not recorded, although they are known to be the same. At 300Hz, the single woofer near-field response as plotted is 17.2dB above the 1m response. Fortunately, we know the response must be continuous across the 300Hz boundary so that we need only drop the near-field woofer data by 17.2dB to join the two curves. But what about the port response?

We'll now look at still another approach to equalizing port and woofer near-field responses. *Figure 5.25a* shows the port response at 50Hz to be roughly 9dB above the single woofer level at 300Hz, but there are two woofers in this example. How do we equalize the port response relative to the woofer pair? With multiple woofers or ports, the easiest way to do this is with area ratios. Area ratios are proportional to the *square* of the diameter ratios. The port area is 20.3cm^2. The manufacturer gives 87cm^2 for S_D. With these values the port correction term is:

$$\text{correction} = 20\log_{10}\left(\frac{S_p}{2S_D}\right)^{\frac{1}{2}} = 10\log_{10}\left(\frac{20.3}{174}\right) = -9.3\text{dB}$$

where I have used $2S_D$ in the denominator to account for the fact there are two woofers in this system. The total port correction is then $-(17.2 + 9.3)$ or -26.5dB.

Corrected port and woofer pair responses are plotted in *Fig. 5.25b* together with the tweeter response. The port response indicates that the woofer pair in this enclosure is flat to 50Hz. The woofer pair spreading loss levels out at 92dB below 300Hz and sets the overall system sensitivity level. Woofer-pair response extends out to almost 10kHz with a broad peak at 1kHz of roughly 4dB relative to the response level at 3kHz and beyond. Tweeter response rises at 12dB/octave with increasing frequency below 2kHz, averaging 94dB above this frequency.

One design goal for this system was to use as low a crossover frequency as practical to ensure wide, uniform polar response. The tweeter rolloff below 2kHz sets an absolute lower limit on the crossover frequency. I chose a fourth-order Linkwitz-Riley acoustic crossover at 2500Hz as the target design. The final system response and the individual driver responses are shown on an expanded scale in *Fig. 5.26*.

Response below 200Hz is the same as that in *Fig. 5.25b*. The 1/3-octave response is flat within ±1.1dB over the range of 200Hz to 20kHz. Crossover occurs at 2.4kHz. System impedance is shown in *Fig. 5.27*. Because the 8Ω woofers are wired in parallel, the system nominal impedance

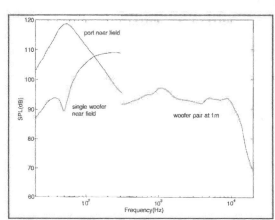

FIGURE 5.25a: Woofer and port responses of Example 5.3.

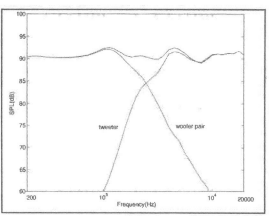

FIGURE 5.26: System and driver responses at 1m for Example 5.3.

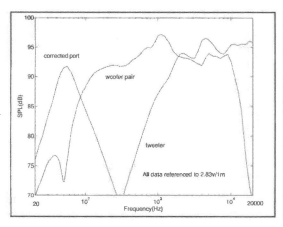

FIGURE 5.25b: Example 5.3's woofer pair, port, and tweeter responses.

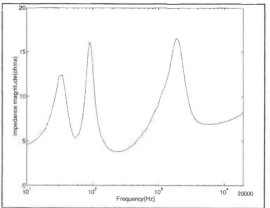

FIGURE 5.27: Example 5.3's system impedance plot.

is 4Ω. As in the previous examples, the impedance rise at 2kHz is a parallel resonance caused
by the interaction of woofer and tweeter crossover
networks.

5.3.3.2 POLAR RESPONSE

Normalized horizontal polar responses are plotted
in *Fig. 5.28* for angles of ±15°, ±30°, and ±45° off-
axis. Horizontal polar response is symmetric about
the axis of rotation. Below 10kHz the total variation over a 90° horizontal angle is within 3.5dB of
the on-axis response. Below 9kHz the figure is
2dB! This is excellent performance. We see the
expected dip below 2kHz followed by the rise just
above 2kHz as acoustic output transitions from
the woofer pair to the tweeter. But even at 45° the
dip-to-peak rise is just 3.3dB.

Figure 5.29 is a plot of normalized vertical polar
responses for the MTM system. Response curves
are shown for ±5°, ±10°, and ±15°. The MTM
geometry forces the vertical polar response to be
symmetric about the tweeter axis. The ±5° curve is
within 0.5dB of the on-axis response! This means
that a frequency response over a 10° arc centered
on the tweeter is virtually indistinguishable from
the on-axis response.

At a typical listening distance of 3.5m, there
exists a window 60cm high (about 2') where
response is very uniform. This broad range accom-
modates seated listeners of all heights. The worst-
case departure of the ±10° curve from the on-axis
response is only 1.8dB.

5.3.3.3 VERTICAL POLAR RESPONSE AND SLOW-SLOPE CROSSOVERS

Although not a part of this example, this is a good
place to talk about the effect of slow-slope
crossovers on polar response. All of the two-way
examples so far in this chapter have used fourth-
order acoustic crossovers. Slow-slope crossovers
and, in particular, 6dB/octave crossovers, are popular with many designers and audiophiles. When
properly implemented, 6dB/octave crossovers can
produce a system with flat frequency response and
low phase shift on at least one axis. Such systems
are often referred to as "transient perfect."

As I have said countless times in the past, however, in loudspeaker system design there is no free
lunch. The slow out-of-band attenuation of a
6dB/octave network means that driver outputs will
overlap significantly around the crossover frequency. This in turn gives a broad region where driver
interference can occur, a region often as large as two
octaves above and below the crossover frequency.

Figure 5.30 shows the vertical polar response of
an MTM system similar to that of Example 5.3.
However, this system crosses over at 3kHz with a
6dB/octave network. All drivers in this system are
mounted flush on the baffle, placing the acoustic
centers of the midbass drivers about 2cm behind
the tweeter acoustic center. This leads to a time
delay of about 58μs and a linear phase shift with
frequency. At 3kHz the phase shift is 60°. (I will discuss the relationship between time delay and phase
shift in Chapter 6.)

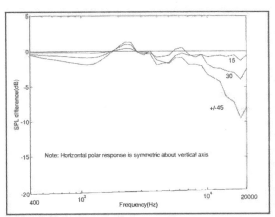

FIGURE 5.28: Example 5.3's normalized horizontal polar response.

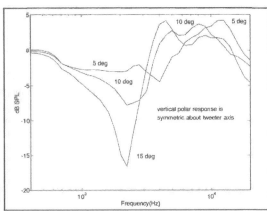

FIGURE 5.30: Vertical polar response of MTM system with 6dB/octave crossover.

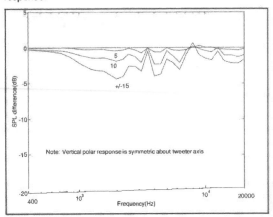

FIGURE 5.29: Example 5.3's normalized vertical polar response.

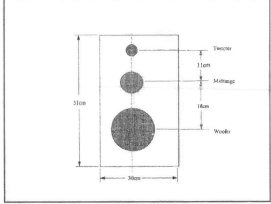

FIGURE 5.31: Baffle and driver layout for Example 5.4.

In addition to the 6dB/octave network, the designer of this system added an all-pass phase shift network to lag the tweeter by 60° at 3kHz. Unfortunately, a single all-pass network is a poor approximation to a time delay. The phase was correct at 3kHz, but incorrect just about everywhere else.

Compare *Figs. 5.30* and *5.29*. The ±5° curve in *Fig. 5.30* averages 4dB below the on-axis response from 1kHz to 3kHz and 4dB above on-axis above the crossover frequency. The ±10° curve dips 7dB at 2kHz and peaks 4dB at 4.5kHz. The ±15° curve is a nightmare. If the tweeter is placed at normal seated listening height, upon standing, this system will sound very bright with a sucked-out midrange. Even though this system is relatively flat on-axis, severe colorations in sound will be heard with small changes in listener height.

5.4 EXAMPLE 5.4: A THREE-WAY, THREE DRIVER SYSTEM

The three-way system to be examined here consists of a 220mm woofer, a 77mm midrange driver, and an 11mm tweeter. The woofer has a polykevlar sandwich cone with a half-roll Neoprene surround mounted in a cast aluminum frame. The 77mm midrange unit and the 11mm tweeter employ concave microthin ceramic diaphragms.

Baffle dimensions and driver spacing for this example are shown in *Fig. 5.31*. The drivers are

aligned vertically with their centerline offset to spread and reduce the effects of edge diffraction. A pair of these speakers must be built in mirror image. System crossover frequencies are fourth-order acoustic at 400Hz and 3kHz. The system is ported.

System impedance taken with the voltage-divider method is shown in *Fig. 5.32*. This plot does not look like the classic vented-system curve, but the woofer near-field response shown in *Fig. 5.33* clearly indicates vented-system response with an f_B of approximately 35Hz. With the woofer crossover network removed, the vented-box upper impedance peak rises to over 75Ω at 80Hz. This unloads the woofer crossover and causes a 3–4dB bass emphasis centered on 100Hz.

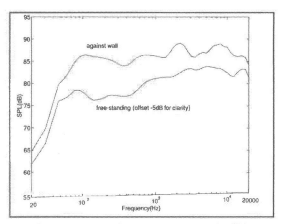

FIGURE 5.34: Full range frequency response of Example 5.4.

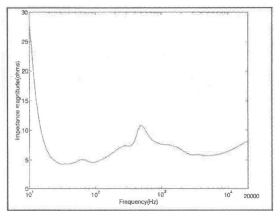

FIGURE 5.32: Example 5.4's impedance plot.

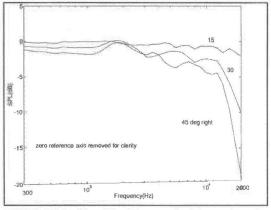

FIGURE 5.35a: Example 5.4's normalized horizontal polar response to right side.

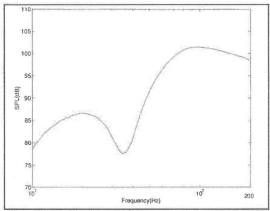

FIGURE 5.33: Woofer near-field response of Example 5.4.

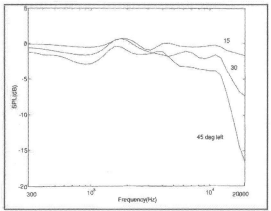

FIGURE 5.35b: Example 5.4's normalized horizontal polar response to left side.

A series RLC shunt tuned to 80Hz was placed across the woofer terminals to suppress the upper impedance peak and thereby level the bass response. If you examine the plot carefully, you can still see a small bump in the impedance curve at 80Hz. The rapid impedance rise below 10Hz is actually the upper side of the lower impedance peak. Its maximum value is below 10Hz.

Full-range frequency response for this example is shown in *Fig. 5.34*. This is a combination of 1m data above 300Hz combined with ground-plane data below 300Hz. You are quite familiar with this kind of procedure now, so the details will not be given.

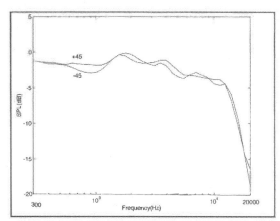

FIGURE 5.35c: Comparison of normalized horizontal polar response at ±45°.

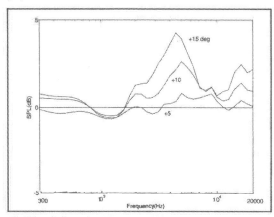

FIGURE 5.36a: Example 5.4's normalized vertical polar response above tweeter axis.

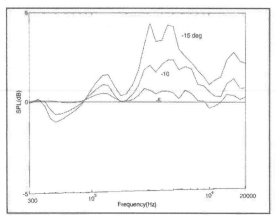

FIGURE 5.36b: Example 5.4's normalized vertical polar response below tweeter axis.

This system was designed to be placed near a wall. *Figure 5.34* shows both freestanding response and response when placed against a single wall. The lift provided by the wall is apparent in the plots. The wall placement lifts the lower end of the spectrum, but it also induces a few extra bumps in the response.

Polar-response data was taken with the system in the freestanding position. Because the drivers are offset relative to the baffle centerline, there is the possibility for a nonsymmetric horizontal polar response. Horizontal response to the right when facing the speaker is plotted in *Fig. 5.35a* for angles of 15°, 30°, and 45°. The 15° line is within 0.5dB of flat at 10kHz and down only 2dB at 20kHz. The 30° and 45° lines are down 2.5dB and 4.5dB, respectively, at 10kHz.

The corresponding curves for polar response to the left side are shown in *Fig. 5.35b*. The curves are similar in shape to those of the right side with just a bit larger dip in the 45° curve below 1kHz. This is more readily seen in *Fig. 5.35c*, where the left and right 45° curves are plotted together. Aside from this difference, they track one another closely.

Vertical polar responses above and below the tweeter axis are shown in *Figs. 5.36a* and *b*, respectively. The expanded vertical scale of these graphs makes the vertical polar response look much worse than it is. All of the departure from flat response occurs in the frequency range around the midrange to tweeter transition. The ±15° curves peak between 3–4kHz. The ±10° curves are plotted together in *Fig. 5.36c*. From the graph we see that the 20° vertical arc falls within a 2.5dB envelope. The ±5° curves are within 0.5dB of the on-axis response.

5.5 EXAMPLE 5.5: A TWO-WAY SYSTEM WITH A LARGE RIBBON DRIVER

All of the examples presented so far have used drivers with dimensions that are small compared to the wavelengths they are called upon to radiate. The system in this example consists of two 30″ ribbon drivers stacked one above the other. They are mounted above a midbass module containing two 5.25″ midbass drivers. According to the manufacturer, crossover from the ribbons to the midbass module occurs at 1kHz. The midbass module rolls off below 100Hz.

This is not a full-range system. It is intended to be used with a subwoofer. The stacked ribbon drivers form a line source 60″ in length. Because the wavelengths they radiate are much shorter than their length, the ribbons behave very differently from more conventional drivers. Before testing the full system, let's look at the special properties of this line source.

For wavelengths much longer than their circumference, conventional circular piston drivers act like point sources in the far-field, producing a spherical radiation pattern. As we have seen, under this condition, sound pressure level falls off by 6dB for each doubling of distance from the driver.

Contrast this behavior with a *finite length* line source. For wavelengths much shorter than its length, a line source's polar response is cylindrical. SPL falls only 3dB for each doubling of distance in the near field. Since the far field is not reached until distances on the order of 15–20 times the line length, you are generally listening to this driver in its near field.

Some simple tests will illustrate the polar-response properties of these ribbon drivers. *Figure 5.37* shows the microphone locations used during these tests. *Figure 5.38* plots the results of a set of three frequency-response tests that illustrate the cylindrical polar-response pattern of a single 30″ ribbon. For these tests the microphone was placed in three locations: on the ribbon centerline, at the ribbon's top edge, and 10″ above the top edge, all at a distance of 2m.

Concentrate on frequencies above 2kHz, where the 30″ ribbon is at least five times longer than a wavelength. Relative to the response on the centerline, response at the upper edge averages 5dB less across all frequencies above 2kHz. Ten inches above the upper edge response is down 15dB on average. You see that SPL levels in the vertical direction fall rapidly beyond the ribbon's edge.

Figure 5.39 shows the 30″ ribbon response on its centerline at one and two meters. The 2m response curve averages 3.3dB less across all frequencies above 2kHz. Together, the results presented in *Figs. 5.38* and *5.39* show that the 30″ ribbon driver produces an essentially cylindrical sound field 30″ high.

In multiway systems using conventional drivers,

all drivers act as point sources in the far field. If SPL levels of the individual drivers match at one distance, they will match at any other distance because they all fall off at the same rate of 6dB for each doubling of distance.

Hybrid systems using drivers with different distance laws present a problem. In the absence of reflections, driver levels in a hybrid system will match at only one location. Moving toward the system from this location, the conventional driver level will rise above the line source level. Moving away, the opposite is true. Room reflections may modify this behavior somewhat, but the direct field, which is primarily responsible for imaging, will behave pretty much as I have described it.

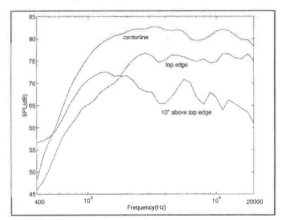

FIGURE 5.38: Illustrating polar-response pattern of ribbon driver of Example 5.5.

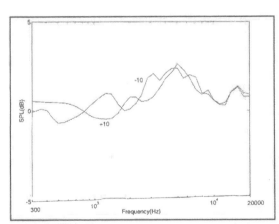

FIGURE 5.36c: Comparison of 10° vertical responses of Example 5.4.

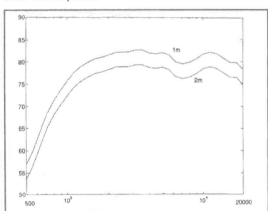

FIGURE 5.39: Illustrating ribbon distance law of Example 5.5.

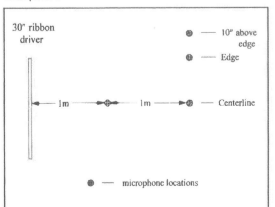

FIGURE 5.37: Microphone locations for examining ribbon polar pattern.

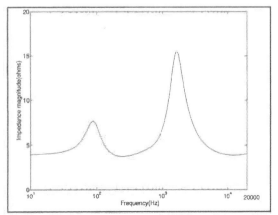

FIGURE 5.40: Example 5.5's ribbon-system impedance.

Let's move on to the system tests. System impedance is plotted in *Fig. 5.40*. The first impedance peak of 7.7Ω at 89Hz indicates the closed-box resonant frequency of the midbass driver pair. The minimum system impedance occurs at 240Hz. Its value there is 3.7Ω. The ribbons plus their crossover capacitor interact with the midbass drivers and their crossover network to form a parallel resonant condition at 1.68kHz, where the impedance peaks to 15.5Ω. The impedance magnitude returns to the 4Ω level above 10kHz.

Next, I ran a trial frequency-response test to determine proper phasing for the ribbons. The manufacturer indicated that polarity reversal may be necessary for smoothest response. The microphone was placed at a distance of 2m and a height of 38" for these tests and most of the remaining tests. *Figure 5.41* shows system response with the stacked ribbon pair connected in phase and with reverse polarity. The "in-phase" condition produced a sharp 26dB response dip at 1.7kHz. Reversing the polarity produced a very smooth transition between the midbass module and the ribbons.

Figure 5.42 is a plot of the system, ribbon, and midbass module frequency responses at 2m. These curves are a combination of 2m free-field responses taken above 400Hz spliced to ground-plane data below 400Hz. Because of the differing distance laws, this data cannot be easily extrapolated back to 1m for sensitivity evaluation. I'll get to that in a moment. Although the electrical crossover appears to be set at 1kHz, the different mid-bass and ribbon SPL levels push the acoustic crossover out to 1.7kHz.

Examining *Fig. 5.42* further, we see that the average level of the midbass drivers between 200 and 800Hz is about 82dB. Relative to this level, the midbass module is down 3dB at 80Hz. The average level of the ribbons between 3–10kHz is roughly 76dB. Between 1–3kHz the system shelves down 6dB. Based on their differing distance laws, moving out to 4m would reduce the shelving to 3dB. An 8m distance (about 26') is needed to bring the system to flat response!

We must be clear here that I am talking about only the on-axis direct-field response. At normal listening distances of 3–4m, the first arrivals will be deficient in treble energy. As you will see in a moment, the system's horizontal polar response is very broad so that lateral reflections will add to the total sound field. Whether or not this system sounds treble deficient is very much a function of the listening environment.

Figure 5.43 shows the ground-plane response of the midbass module with its crossover. The microphone was placed at 1m and response data was then corrected to 1m free-field. This curve places the midbass module sensitivity at 92dB SPL/2.83V/1m. This agrees with manufacturer's specifications, but the shelved response makes assignment of a single sensitivity number to the entire system arbitrary.

Figure 5.44 is very interesting. This plot com-

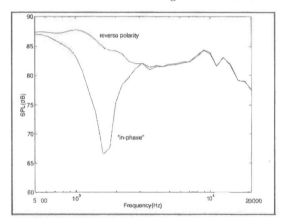

FIGURE 5.41: Example 5.5's ribbon polarity test.

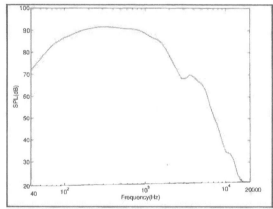

FIGURE 5.43: Example 5.5's midbass module ground-plane response.

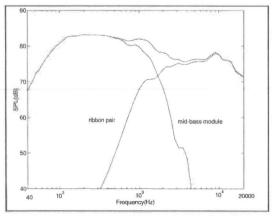

FIGURE 5.42: Ribbon system and driver responses of Example 5.5.

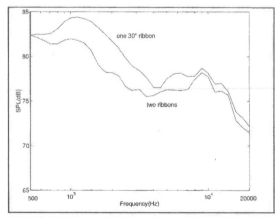

FIGURE 5.44: Example 5.5's system response with one and two ribbons.

pares the system response in its normal configuration using a stacked ribbon pair against the response with only the lower ribbon active. In the transition region between the lower- and upper-frequency system response shelves, the upper ribbon actually *reduces* system on-axis response at the microphone height. Above 4kHz the response is the same whether one or two ribbons are active.

This is just another manifestation of the cylindrical polar response. The 30″ ribbon creates a cylindrical sound field 30″ high. Adding the second ribbon extends the height of the cylindrical field to 60″, but it does not increase the SPL at the microphone position.

We already know that vertical polar response is very uniform within the vertical span of the ribbon. *Figure 5.45* shows response at 30° and 45° off-axis in the horizontal plane. These plots are referenced to the on-axis response. At 15kHz the ribbons are down 3dB and 6dB at 30° and 45°, respectively, relative to the on-axis response. This performance should be compared to that of a typical 28mm soft-dome tweeter. The corresponding numbers are −5dB and −17dB. Horizontal coverage of the ribbons is clearly much broader.

All of the response data shown so far has been essentially anechoic. *Figure 5.46* gives us another look at this hybrid system. This is a 1/3-octave RTA analysis with the microphone placed at a typical listening position, 3m out at a 38″ height. This response contains all room reflections and standing-wave effects in my test area. The general shape of the curve is similar to that shown in *Fig. 5.42*.

5.6 EXAMPLE 5.6: A 12″ SUBWOOFER

The subwoofer in this example consists of a 12″ long-throw woofer with a polypropylene cone and a half-roll rubber surround mounted in an enclosure of 70 ltr net internal volume. The enclosure is vented with two 3″ ID ports that tune the box to 28Hz.

A high-resolution impedance plot of the subwoofer is shown in *Fig. 5.47*. In addition to the double-humped characteristic of a vented system, we see two minor bumps at 200 and 380Hz. The cause of these bumps will be seen in *Fig. 5.48*.

This figure shows the near-field responses of a *single* port and the woofer out to its valid near-field f_{MAX} of 400Hz. The sharp dip in woofer output at 28Hz reveals the box-tuning frequency. Our first job is to find the proper level for the port output relative to the woofer. First recognize that the measured output from two ports will be 6dB greater than the single port curve. Next we must correct the port output by the ratio of total port area to woofer cone area. The area of a single 3″ ID port is 7.07 in^2, or 45.6cm^2, and the woofer cone area is 530cm^2. With these area values, this correction is:

$$\text{area correction} = 10\log_{10}\left(\frac{2 \times 45.6}{530}\right) = -7.6\text{dB}$$

The total correction is then:
$$\text{total correction} = +6\text{dB} - 7.6\text{dB} = -1.6\text{dB}$$

The corrected port pair output is also plotted in *Fig. 5.48*. At 28Hz the combined port output is 25dB above the woofer output. The 28Hz port output is also 4dB below the average near-field response of the subwoofer above 100Hz. Thus the −3dB point of the subwoofer's half-space response is somewhat above 28Hz.

The ripples in port response at 180 and 250Hz are caused by stiffening baffles within the enclosure, creating multiple chamber resonances. We have already seen this effect in Chapter 4. The sharp dip in port response at 320Hz is caused by the first standing-wave mode of the enclosure. Although not shown in *Fig. 5.48*, there are additional dips in port response above 600Hz caused by "organ pipe" resonances of the port tubes. These resonances cause small dips in the woofer's near-field response and show up as bumps on the impedance curve. Fortunately, in its application, the subwoofer is crossed over at 80Hz.

5.6.1 SUBWOOFER POWER RESPONSE

The near-field response confirms the subwoofer design, but doesn't tell us much about how this subwoofer will sound in a typical listening room. To do that, let's look at subwoofer power response. What is power response? As commonly used, power response is just the normal frequency response averaged over several listening positions. For example, a typical home-theater system might have a couch and flanking chairs facing the screen at a 4m distance. This seating arrangement can easily span 3m horizontally.

The frequency response of the loudspeaker system at each location along this span will undoubtedly vary, but the average response over the span should be representative of what the average listener will hear. This average can be made by visualizing an imaginary grid 3m across and perhaps 1m high spanning the listening positions. To get the power response a microphone is placed at each point on the grid and the frequency response recorded. The individual data sets are then averaged to obtain the power response.

Some manufacturers of loudspeakers for sound reinforcement applications publish a power response which is the integrated SPL over a portion of a sphere enclosing the loudspeaker. The measuring patch includes the area inside the −3dB contour. Many designers and loudspeaker system reviewers consider the 1/3-octave averaged power response an excellent measure of loudspeaker performance.[4]

The test area and microphone locations used to measure this subwoofer's power response are shown in *Fig. 5.49*. The subwoofer was placed in one corner of the test area. Since the subwoofer should be omnidirectional over its intended frequency range, I simply placed the microphone in five locations along an arc 8′ from the subwoofer. These are not necessarily the best microphone locations. They should simply be taken as representative of the process. In practice, for full-range systems, many more locations should be used.

Subwoofer RTA responses for the five locations are plotted in *Fig. 5.50*. Each plot is numbered with its microphone position, and they are offset by 15dB for clarity. As you would expect, the greatest variation in response occurs below 200Hz where room modes are widely spread in frequency. The response at position 4 is the smoothest. The average of the five response curves is shown in *Fig. 5.51*. Below 100Hz, peak-to-trough variations of 8dB are seen. There is substantial output in the 25Hz 1/3-octave band and sensible output even at 20Hz, probably due to room lift.

5.7 SUMMARY REMARKS

In this chapter we have applied the measurement principles of Chapter 4 to the challenge of measuring loudspeaker system acoustic response. The examples given in this chapter have come from my files of system designs and tests accumulated over a period of several years. Each example has highlighted some important aspect of the measurement process and given practical advice on how to interpret and use the results.

You have seen several techniques for combining near-field low-frequency data with freestanding or ground-plane measurements. I have discussed the relationship between electrical and acoustic measurements and the crossover design process. We have examined the polar response of several systems and noted its influ-

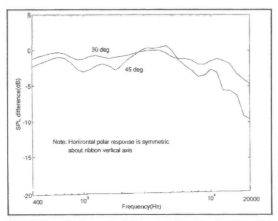

FIGURE 5.45: Example 5.5's normalized horizontal polar response of ribbon system.

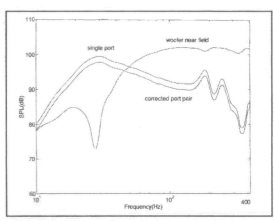

FIGURE 5.48: Subwoofer nearfield driver and port responses of Example 5.6.

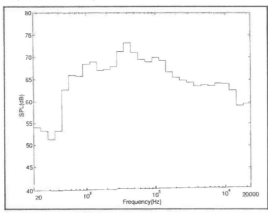

FIGURE 5.46: Example 5.5's RTA response.

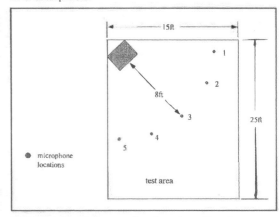

FIGURE 5.49: Microphone locations for subwoofer power-response measurement.

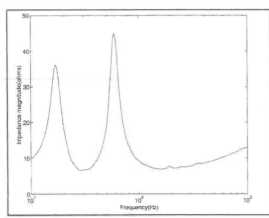

FIGURE 5.47: Subwoofer impedance plot of Example 5.6.

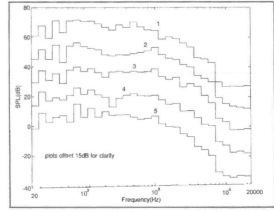

FIGURE 5.50: Example 5.6's RTA response at five locations.

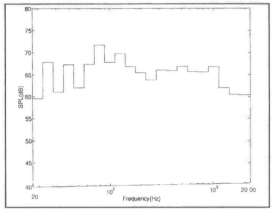

FIGURE 5.51: Subwoofer power response of Example 5.6.

ence on sound quality. One very interesting example treated a large linear array, detailing its characteristics and the impact they have on the measurement process. Given the very simple instrumentation used in acquiring the data presented in this chapter, the amount of system information gleaned from it is really quite amazing.

REFERENCES

1. Struck and Temme, "Simulated Free Field Measurements," *JAES*, Vol. 42, No. 6, pp. 467–482 (June 1994).

2. Dick Crawford, "The Warbler," *The Audio Amateur* 1/79, pp. 22–27, 49.

3. Chapter 2, "Psychoacoustics," *Handbook for Sound Engineers*, second edition, G.M. Ballou, Editor, *SAMS* Division, Macmillon Publishing, Carmel, IN, 1991.

4. J.A. Atkinson, "Loudspeakers: What Measurements Can Tell Us—and What They Can't Tell Us!," 103rd AES Convention, New York, September 26–29, 1997, Preprint No. 4608 (O–5).

5. "A Geometric Approach to Eliminating Lobing Error in Multiway Loudspeaker Systems," *74th Convention of the Audio Engineering Society*, Preprint No. 2000, New York, NY, October 1983.

6. "A High-Power Satellite Speaker," *Speaker Builder*, Issue 4, 1984, pp. 7–14.

7. D'Appolito, Joseph, and Bock, James, "The Swan IV Speaker System," *Speaker Builder*, Issue 4, 1988, pp. 9–21.

CHAPTER 6

TIME, FREQUENCY, AND THE FOURIER TRANSFORM

6.0 INTRODUCTION

Explaining Fourier theory with few, if any, equations is a major challenge. This will also be a difficult chapter for many of you to read. It is very important, however, to have a strong *qualitative* appreciation for the material in this chapter if you are to understand the operation of PC-based acoustic data-acquisition systems and use the wealth of new information that they provide to maximum advantage.

The flowchart in *Fig. 6.1* will familiarize you with the topics in this chapter and the order in which they will be introduced. This should help you to see where we are going with this material and keep you from getting lost in the middle of the discussion. A summary of the material covered in each section follows.

Up to this point, all of the measurement techniques I have described and all of the examples we have examined involve *analog* measurements in the *frequency* domain. All of the graphs presented plot impedance or response versus frequency. The PC data-acquisition systems I will discuss in Chapter 7 work in the time domain. They are the digital counterparts of an oscilloscope, which displays waveforms on a *time* axis.

Fortunately, for *linear* systems, there is a direct relationship between time-domain data and frequency-domain data. For a loudspeaker, the equivalent of its frequency response in the frequency domain is its *impulse response* in the time domain.

Impulse response and frequency response are mathematical duals of one another related by the direct and inverse Fourier transforms. The Fourier transform (FT) is a mathematical operation which transforms time-domain data into frequency-domain data. The transform is *invertable;* that is, if you know one you know the other. This invertability is illustrated below:

Fourier transform

Impulse response ⇄ **Frequency response**

Inverse Fourier Transform

The diagram shows that the FT transforms time-domain impulse-response data into frequency response. Similarly, the Inverse Fourier transform (IFT) takes frequency response and transforms it into the time-domain impulse response.

Section 6.1 introduces the impulse signal and defines a loudspeaker's impulse response with an example. There is a problem, however. Loudspeaker impulse response is a continuous-time function, and frequency response is a continuous-frequency function. By this I mean that the impulse response has a value for every point in time, and the frequency response has a value for every point

in frequency.

But computers, no matter how fast, require a finite amount of time to complete any logical or

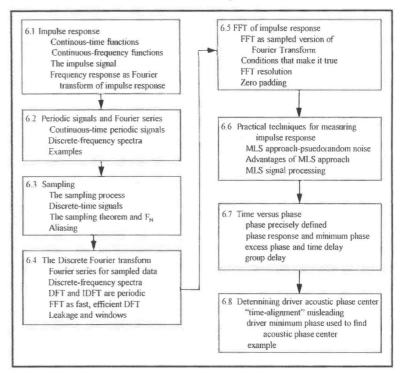

FIGURE 6.1: Chapter 6 outline.

mathematical operation. By themselves, they can neither measure nor generate continuous-time or continuous-frequency functions. They must work with *samples* of either process, known only at discrete instants in time or discrete points in frequency. The next five sections of this chapter explain how we get a good representation of impulse and frequency responses from sampled data.

Section 6.2 defines continuous-time periodic signals and provides examples. The Fourier series representation of these signals is then presented. This series comprises an infinite sum of harmonically related sinusoidal signals. The conclusion from this representation is that a periodic signal has a *discrete-frequency* spectrum.

Section 6.3 describes the sampling process in which a continuous-time signal is converted into a *discrete-time* signal, whose value is known only at discrete, equally spaced instants of time. Next is the sampling theorem, which sets an upper bound on the frequencies which can be unambiguously recovered by the sampling process. I then explain with examples the problem of frequency aliasing.

In Section 6.4 I present the Discrete Fourier Transform (DFT), which is basically a Fourier series for sampled data. Both the DFT and its inverse are periodic. As this section explains, this leads to the problem of frequency leakage when the time-sampled sequence itself is not periodic over the sample interval. I discuss and illustrate with examples a partial solution to leakage using data windows.

The Fast Fourier Transform (FFT) is a computationally efficient algorithm for computing a restricted version of the DFT. Section 6.5 discusses the FFT. Under the right conditions (which are explained), the FFT of a sampled impulse response is shown to be a sampled version of the continuous-frequency frequency response. This section discusses FFT resolution and describes a technique for producing smoother FFT plots with zero padding.

Section 6.6 introduces a practical technique for measuring loudspeaker impulse response using a Maximum Length Sequence (MLS) pseudo-random noise. The MLS signal is described, and advantages of the approach are listed. A block diagram overview of the MLS signal processing is also presented.

Phase is much talked about but little understood. Section 6.7 precisely defines phase, as well as phase response. Examples show that phase angles in excess of a few hundred degrees are possible in many networks and loudspeaker systems. The periodicity of sine waves in phase, however, limits phase-response measurements to ±180°.

This section also explains with an example the process of unwrapping measured phase, and introduces the concepts of a minimum phase response and excess phase. This leads to descriptions of all-pass filters and the phase response of a pure time delay. Group delay is also defined and examples presented.

The discussion of phase in Section 6.7 helps to define the acoustic phase center of a driver in Section 6.8. Then, using the minimum phase property of drivers, the section illustrates a technique for finding the acoustic phase center with a hypothetical example.

6.1 THE IMPULSE RESPONSE

In Chapters 4 and 5 I used a sine-wave oscillator with constant output voltage to plot a point-by-point frequency response—a tedious procedure at best. Suppose we could find one signal that had a constant, flat spectrum at all frequencies. If we fed this signal into a loudspeaker and were then able to perform some kind of high-resolution frequency analysis on the acoustic output, that analysis would yield the loudspeaker's frequency response.

One such signal is the unit impulse—also called the Dirac delta function. This very unusual signal has infinite amplitude and zero width in time, with a fixed area under the impulse of one, hence the modifier, "unit." One way to think of the unit-impulse function is to visualize a rectan-

gular pulse of height, h, and width, w, with a fixed area under the pulse of $h \times w = 1$.

Now let w shrink to zero while maintaining the area fixed at 1 (mathematicians do this all the time). In the limit of the shrinking process, the amplitude of h becomes arbitrarily large to achieve an impulse. The FT of an impulse is a flat spectrum containing all frequencies from DC to infinity with constant amplitude and zero phase at each frequency. If you apply an impulse to a system, the system is excited by all frequencies at the same time.

Recall the damped spring-mass model for the mechanical portion of a driver presented in Fig. 2.3 of Chapter 2. If this system is initially at rest and you strike the mass with an impulse at time zero, the mass will move from its rest position with a damped sinusoidal motion. This motion is illustrated in *Fig. 6.2*. In this example, the mass-spring resonant frequency is set at 40Hz with a mechanical Q of 3. The motion damps out in 5 to 6 cycles, returning essentially to its rest position in about 200ms.

The impulse response is customarily denoted by the symbol $h(t)$. The following equation defines the impulse response shown in *Fig. 6.2*:

$$h(t) = e^{-\left(\frac{\pi f_s t}{Q_M}\right)} \times \sin\left[2\pi f_s\left(1 - \frac{1}{4Q_M^2}\right)^{\frac{1}{2}} t\right] \quad t \geq 0 \quad [6.1a]$$

$$h(t) = 0 \quad t < 0 \quad [6.1b]$$

where:

$$f_S = 40Hz, Q_M = 3 \text{ and } e = 2.71828.......$$

e is the base for the natural logarithm. This is the equation for an exponentially damped sinusoid. It is defined for all real values of time, and is, therefore, a continuous-time function.

The above equations assume that the impulse is applied at time zero. The impulse response has a value for all positive time, but no response before application of the impulsive input. Notice that the frequency of oscillation is shifted downwards somewhat by the damping. For $Q_M = 3$:

$$f_s\left(1 - \frac{1}{4Q_M^2}\right)^{\frac{1}{2}} = f_s\left(1 - \frac{1}{4 \times 3^2}\right)^{\frac{1}{2}} = f_s\left(1 - \frac{1}{36}\right)^{\frac{1}{2}} = 0.986 f_s \quad [6.2]$$

The FT of the impulse response is a complex frequency-domain function. It has both real and imaginary parts, much like an impedance function. The *magnitude* of the transformed impulse response is the system frequency response. The *angle* of the transformed impulse response is the phase response.

By synchronizing the start time for recording an impulse response to the time of the impulse itself, you get both amplitude and phase information. The FT of the impulse response is commonly given the symbol H(f), where f denotes that the transform is a function of frequency.

If you calculate the Fourier transform for the impulse response of *Fig. 6.2*, you get the follow-

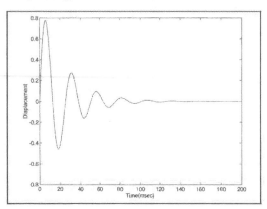

FIGURE 6.2: Impulse response of damped spring-mass system.

ing magnitude and phase responses:

$$\left|H(f_n)\right| = \frac{1}{\left[f_n^4 - \left(\frac{17}{9}\right)f_n^2 + 1\right]^{1/2}} \quad f_n \geq 0 \quad [6.3a]$$

and

$$\text{phase}[H(f_n)] = -\left\{90 + \arctan\left[\frac{3\left(f_n^2 - 1\right)}{f_n}\right]\right\} \text{ deg } [6.3b]$$

where:

$$f_n = f/f_S$$

The f_n expression is a normalized frequency, and it is a convenient substitute for the more cumbersome expression (f/f_S). At resonance, f_n equals 1.

The arc-tangent function above is the inverse of the trigonometric tangent. For example, the tangent of 45° is one. If you ask what angle has a tangent equal to one?—the answer is 45°. This is shown below in equation form:

If $\tan(45°) = 1$, then $\arctan(1) = 45°$

The magnitude and phase functions are shown in *Figs. 6.3* and *6.4*, respectively. The magnitude response is plotted in dB. *Figure 6.3* should be compared with the family of frequency-response curves shown in Fig. 2.6a. Both magnitude and phase are

defined for all possible values of frequency. They are called continuous-frequency functions. Notice that the magnitude response plot peaks just slightly below 40Hz. This corresponds to the shifted frequency computed in Equation 6.2.

At zero frequency or DC, $f_n = 0$, $|H(0)|$ is one or 0dB and the argument of the arc-tangent function is negative infinity. This corresponds to an angle of −90°, which just cancels the +90° term within the outermost brackets of Equation 6.3b. Thus at DC the mass moves one-for-one with the input.

The phase angle is negative for all frequencies above zero. At $f = f_S$, $f_n = 1$, the argument of the arc-tangent function is zero and thus the phase response of the spring-mass system is −90°. Above resonance, phase grows increasingly negative, until it settles out at its final value of −180° well above f_S.

Together, the frequency and phase-response pair are called the loudspeaker *transfer function*. The loudspeaker takes an input voltage and "transfers" it to an output SPL at a fixed distance assuming a constant input voltage of 2.83V. The units of this transfer function are dB SPL/2.83V at 1m. I used this nomenclature throughout Chapters 4 and 5.

The true unit impulse is a mathematical idealization which cannot be realized in practice. In practice you could use a very narrow pulse. For example, a 6µs wide rectangular pulse has a spectrum that is down only 1dB at 20kHz. Indeed, some of the earlier time-domain approaches to loudspeaker response analysis used a rectangular pulse.[1,2]

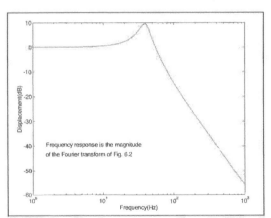

FIGURE 6.3: Displacement frequency response of spring-mass system.

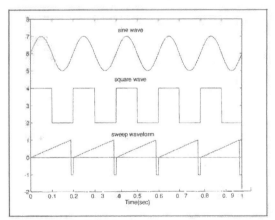

FIGURE 6.5: Periodic signals.

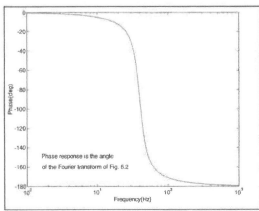

FIGURE 6.4: Displacement phase response of spring-mass system.

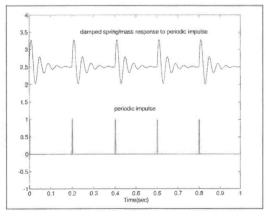

FIGURE 6.6: More periodic signals.

As I indicated in the introduction, however, a problem occurs with trying to use the FT in the manner I have described. The impulse response is a continuous-time function and frequency response is a continuous-frequency function. But computers inherently work with samples of the signal which are known only at discrete instants in time. In the next several sections I will cover aspects of Fourier theory and sampling needed to understand how to compute accurate transfer functions from sampled data.

6.2 PERIODIC SIGNALS AND FOURIER SERIES

In this section I describe periodic signals and show you how to represent them in terms of Fourier series. I touched upon the concepts of period and frequency in connection with sine waves in Chapter 2. I will generalize those ideas here and then make them more concrete with examples.

6.2.1 PERIODIC SIGNALS

Periodic signals are those whose values repeat in time on a periodic basis. The repetition occurs over a *fixed* time interval called the *period* of the signal. *Figure 6.5* depicts three periodic signals. You are already familiar with two of the signals shown—the sine wave and the square wave.

The third signal depicts a typical oscilloscope sweep voltage waveform. As the linearly rising portion of the periodic voltage sweeps the CRT

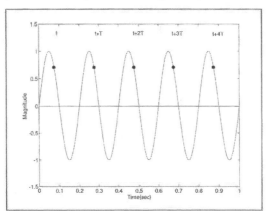

FIGURE 6.7: Illustrating the periodic nature of a sine wave.

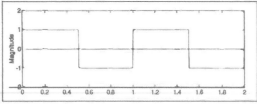

FIGURE 6.8a: Square wave for Equation 6.8.

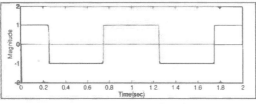

FIGURE 6.8b: Square wave for Equation 6.9.

electron beam across the screen, a time-domain replica of the signal is painted on the screen. The short negative pulse following the ramp returns the beam to its starting position for the next sweep.

Figure 6.6 depicts a very different set of periodic signals. The sequence of damped sinusoids is the displacement response of our spring-mass model to a periodic sequence of impulses. Note that the response of this model to a periodic sequence of impulses is itself periodic. I will refer to this example again later in the chapter.

The symbol, T, from Chapter 2 denotes the period of a sine wave. A period is the time it takes for the sine wave to complete one cycle. Equivalently, it is the time it takes for the sine-wave phase to rotate through an angle of 360°.

Let's use the symbol $x(t)$ to denote an arbitrary signal as a function of time. By this I mean that as time progresses, the value of x changes according to some mathematical definition dependent on time. If $x(t)$ is periodic, then the following is true:

$$x(t) = x(t + T) = x(t + 2T) = x(t + 3T) =$$
$$........= x(t + nT) \qquad [6.4]$$

where n is an integer. *Figure 6.7* illustrates this general property of periodic signals.

You can enter the graph of the sine wave at any point in time, t, and read off its value. At any later time, t + nT, the sine wave will have the same value. Equation 6.4 and *Fig. 6.7* show that if you know the value of the sine wave for every value of time in one cycle or period, you know its value for all other times. This is true of any periodic signal.

6.2.2 FOURIER SERIES

This section covers the Fourier series representation for continuous-time periodic signals. Those of you with some calculus background should not confuse "continuous-time" with the mathematical concept of continuity and continuous functions. Later, when sampling is discussed, you will see signals that are known only at the sample times which are discrete points in time.

The basic mathematical representation of periodic signals is the Fourier series, which is a linearly weighted sum of harmonically related sinusoids. Jean Baptiste Joseph Fourier (1768–1830) used these trigonometric series to describe heat conduction and the temperature distribution in heat-conducting materials. Although the trigonometric series bears Fourier's name, Daniel Bernoulli (1700–1782) introduced the sine-cosine series in 1738 to represent a function of two variables in the solution of the wave equation for a vibrating string.[3]

Bernoulli later asserted that "every function" has a Fourier series representation over the time interval [0,T]. Fourier obviously shared this belief, but many mathematicians of the time disputed it and a controversy ensued over the properties of the Fourier series which was not fully answered until the second half of this century!

6.2.2.1 THE GENERAL FORM OF THE FOURIER SERIES

The Fourier series has the form:

$$x(t) = A_0 + A_1\cos(2\pi f_0 t) + B_1\sin(2\pi f_0 t) + A_2\cos(4\pi f_0 t) + B_2\sin(4\pi f_0 t) + \\ +A_3\cos(6\pi f_0 t) + B_3\sin(6\pi f_0 t) + \ldots \\ +A_k\cos(2\pi k f_0 t) + B_k\sin(2\pi k f_0 t) + \ldots$$

A more compact form of this series is:

$$x(t) = A_0 + \sum_{k=1}^{k=\infty}[A_k\cos(2\pi k f_0 t) + B_k\sin(2\pi k f_0 t)] \quad [6.5]$$

The Σ symbol indicates that a summation of harmonically related terms is to be performed where the frequency index, k, takes on values from 1 to infinity (∞).

The A_0 term represents the average value of the signal x(t) over a period. It is also called the DC value. The f_0 term is the *fundamental frequency* of the signal, where:

$$f_0 = \frac{1}{T} \quad [6.6]$$

The f_0 term is the reciprocal of the period, T, and has the units of cycles per second or Hertz (Hz).

The cosine and sine terms associated with the A_1 and B_1 coefficients are called the fundamental or first harmonic of the signal because they oscillate at the fundamental frequency, f_0. The higher-order terms with frequencies of $2f_0$,

$3f_0, \ldots kf_0$ are referred to as the second, third, and kth harmonics of the signal, respectively.

The coefficients A_k and B_k are computed with the following formulas:

$$A_0 = \frac{1}{T}\int_t^{t+T} x(t)dt$$

$$A_k = \frac{2}{T}\int_t^{t+T} x(t)\cos(2\pi k f_0 t)dt \quad [6.7]$$

$$B_k = \frac{2}{T}\int_t^{t+T} x(t)\sin(2\pi k f_0 t)dt$$

These are presented for the sake of completeness. If you are not comfortable with the integrals in Equations 6.7, don't worry about it. I will not use these equations again.

6.2.2.2 THE FOURIER SERIES OF A SQUARE WAVE

It's time for an example. The square wave shown in *Fig. 6.8a* has a period of one second and therefore a fundamental frequency of one Hertz. If the integrals given in Equations 6.7 are evaluated over the time interval [0,1] seconds, the following Fourier series is obtained:

$$x(t) = \frac{4}{\pi}\{\sin(2\pi \times f_0 t) + \frac{1}{3}\sin(2\pi \times 3f_0 t) + \frac{1}{5}\sin(2\pi \times 5f_0 t) \\ + \frac{1}{7}\sin(2\pi \times 7f_0 t) + \ldots + \frac{1}{2k-1}\sin[2\pi \times (2k-1)f_0 t]\}$$

or, in a more compact form:

$$x(t) = \frac{4}{\pi}\left[\sum_{k=1}^{k=\infty}\left(\frac{1}{2k-1}\right)\sin[2\pi \times (2k-1)f_0 t]\right] \quad [6.8]$$

Notice that the term, $(2k-1)$, is an odd number for all integers k. Thus the series has only sine terms with only odd multiples of the fundamental frequency. I'll say more about this shortly.

Of course, in practice, we cannot compute an infinite number of terms. The question that arises

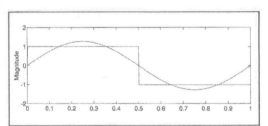

FIGURE 6.9a: Square wave and first harmonic.

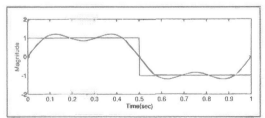

FIGURE 6.9b: Square wave and first two terms of Fourier series.

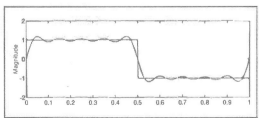

FIGURE 6.9c: Square wave and first five terms of Fourier series.

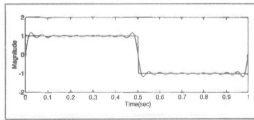

FIGURE 6.9d: Square wave and first ten terms of Fourier series.

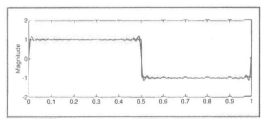

FIGURE 6.9e: Square wave and first 20 terms of Fourier series.

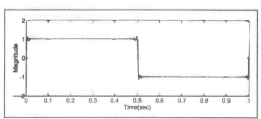

FIGURE 6.9f: Square wave and first 50 terms of Fourier series.

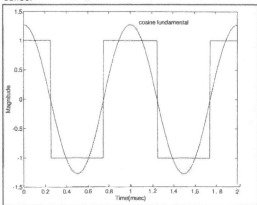

FIGURE 6.10: Time-shifted square wave and its fundamental.

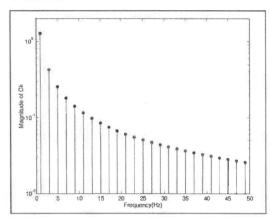

FIGURE 6.11: Square-wave line spectrum.

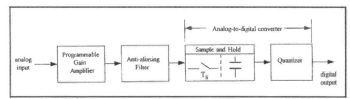

FIGURE 6.12: The sampling process.

is: How many terms are enough? The answer depends on the periodic signal itself and the level of approximation you require.

Figures 6.9a–f illustrate the progressively better approximation to a square wave that occurs as more terms of the Fourier series are added to the summation. *Figure 6.9a* shows the square wave and the first term, the fundamental term, of the series. *Figure 6.9b* compares the square wave against the first two terms of the series. *Figures 6.9c, d, e,* and *f* compare the square wave against the sum of the first 5, 10, 20, and 50 terms of the series, respectively.

If your goal is to get a good representation of the +1 and −1 levels of the square wave, the first

ten terms of the series may be enough. If you wish to capture the sharp +1 to −1 (and the −1 to +1) transition, you may need 50 terms or more. Whatever number of terms you choose, however, you are assured that you have the best least-squares approximation to the square wave for that number of terms.

For example, if you choose to use the first ten terms of the series, there is no other combination of ten sine and cosine terms that will yield a better fit to the periodic function. Furthermore, if you add an 11th term, the fit will always be better; that is, each additional term added to the series always improves the fit.

That said, the *ideal* square wave is a particularly difficult signal to approximate because of the sharp discontinuity. The transition from +1 to −1 occurs in zero time. This discontinuity induces ripples in the Fourier approximation which can be seen in *Fig. 6.9*. These ripples are called the *Gibbs* phenomena.[3,4]

Although the ripple frequency increases with the addition of more terms, the ripple amplitude in the immediate vicinity of the transition does not die away. For a large number of terms, the overshoot at the transition is about 18%. Fortunately, real systems and real square-wave generators have finite bandwidth and are slew-rate-limited. In practice, the transition will occur over a finite time interval, with no discontinuity in the periodic signal and hence little, if any, Gibbs phenomena.

6.2.2.3 FOURIER SERIES AND THE FREQUENCY DOMAIN

Your first impression of the Fourier series might be that it is just an alternate time-domain expression for a periodic signal. But it is much more than that. It gives us a frequency-domain description of the signal, has information on the phase of each component of the series, and is also a data-compression algorithm. Let's first look at the phase information.

In *Fig. 6.8a*, the one-second period of the square wave was taken to be the time interval between two −1 to +1 transitions. This is also the time interval [0,T]. However, the 6.7 equations show that the coefficients can be evaluated over any time interval [t, t + T].

In *Fig. 6.8b* I have shifted the square wave to the left one-quarter of a cycle. The period is still one second, but now it is measured from mid-point to mid-point of the +1 level. The Fourier series for this square wave is:

$$x(t) = \frac{4}{\pi}\left[\sum_{k=1}^{k=\infty}\left(\frac{1}{2k-1}\right)\cos[2\pi \times (2k-1)f_0 t]\right] \quad [6.9]$$

As before, the series contains only odd harmonics, but the terms are all cosine now rather than sine. Since the cosine and sine are 90° apart in phase, shifting the square wave one-quarter cycle to the left in time shifts all harmonics by 90°. Notice that you could have obtained the same result by evaluating the integrals in the 6.7 equations over the time interval [0.5,1.5] seconds in *Fig. 6.9a*.

Fig. 6.10 shows the time-shifted square wave

and its cosine fundamental. If the shift had been less than one-quarter cycle, both sine and cosine terms would have appeared in the Fourier series for the square wave. The A_n and B_n coefficients are highly dependent on where you choose to start the time axis. Can you find a different form of the Fourier series in which this is not true?

Each sine/cosine pair in the Fourier series can be written in an alternate form:

$$A_k \cos(2\pi k f_0 t) + B_k \sin(2\pi k f_0 t) = C_k \sin(2\pi k f_0 t + \theta_k) \qquad [6.10]$$

where:

$$C_k = \sqrt{A_k^2 + B_k^2}$$

and

$$\Theta_k = \arctan\left(\frac{A_k}{B_k}\right)$$

In Equation 6.10, C_k represents the *amplitude* or magnitude of the harmonic, and θ_k is its phase relative to the time we choose to start the period. Now the interesting point here is that regardless of where we choose the value of t which starts the time interval, [t,t + T], C_k will not change. Only θ_k changes to account for the time shift.

This makes sense intuitively. The frequency content of any periodic signal is an intrinsic property of that signal. Changing the starting time of the interval over which the Fourier series is calculated should not change the signal's frequency content. All it should do is change the phase of each harmonic.

The C_k coefficients make up the *frequency magnitude spectrum* of a periodic signal. *Figure 6.11* is a plot of the frequency magnitude spectrum of a 1Hz square wave. The spectrum contains only discrete values starting at 1Hz and spaced 2Hz apart. This type of spectrum is called a discrete-frequency spectrum or a *line* spectrum.

Contrast this spectrum with the frequency-response spectrum of a loudspeaker which has a value at every frequency. The loudspeaker response is a continuous-frequency spectrum. You will see line spectra again in connection with the discussion of sampling and the DFT. Fortunately, under the right conditions, there is a direct relationship between the discrete-frequency and continuous-frequency spectra. This point will be discussed further in Section 6.5.

I mentioned the data-compression property of the Fourier series. Though not central to the goal of loudspeaker testing, this property is interesting in its own right, as it highlights a characteristic of many mathematical transforms. As an example, say you test a woofer for harmonic distortion and the results show that at 30Hz and 95dB SPL the woofer produces significant second, third, fifth, and seventh harmonics of the 30Hz fundamental. The Fourier series for the woofer's acoustic waveform will have ten coefficients, either five As and five Bs or five Cs and five θs.

Two coefficients represent the 30Hz funda-

mental and four additional coefficient pairs represent the distortion products. At this point, ten numbers, the Fourier coefficients, have replaced the entire periodic function of the woofer acoustic output. If at any later time you wish to reproduce the woofer waveform, simply recall the ten coefficients from memory and plug them into Equation 6.5 or Equation 6.10 to regenerate the woofer output.

For band-limited periodic signals, the Fourier series takes a periodic function, x(t), with an infinite number of values in the time domain, and transforms it into a finite set of coefficients in the frequency domain. Once the coefficients are known, they may be used to transform the frequency-domain description back into the time domain. You will see this bidirectional property again in the discussion of the DFT in Section 6.4.

6.3 SAMPLING AND THE SAMPLING THEOREM

This section describes the sampling process and the sampling theorem. Computers are inherently discrete-time data processors. Sampling is the process of taking snapshots of analog data at discrete time increments. The data can then be converted to digital form for further processing in a computer.

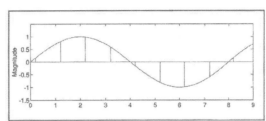

FIGURE 6.13a: 125Hz sine wave sampled at 1kHz.

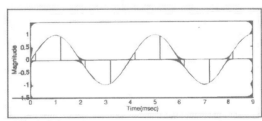

FIGURE 6.13b: 250Hz sine wave sampled at 1kHz.

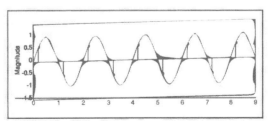

FIGURE 6.13c: 500Hz sine wave sampled at 1kHz.

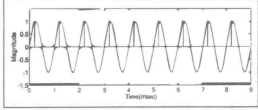

FIGURE 6.13d: 1000Hz sine wave sampled at 1kHz.

6.3.1 THE SAMPLING PROCESS

The sampling process is illustrated in *Figure 6.12*. A typical PC data-acquisition card contains a programmable gain amplifier, an anti-aliasing filter, and an analog-to-digital (A-D) converter. Integrated circuit A-D converters usually have a fixed analog input range. Common ranges are 0–5V or –2.5 to +2.5V.

The programmable gain amplifier (PGA) adjusts the level of the input signal to get the maximum dynamic range from the A-D converter. For example, if a 1V pk-pk sine wave is input directly into a 12-bit converter with a 5V input range, the first two most significant bits (MSB) would be zero and the sine wave would effectively be quantized to only ten bits. (The first two MSBs divide the full scale range of the A-D converter by four for a 1.25V range.) By setting the PGA gain to 5, the amplified 1V input signal will now span the full 12-bit range of the A-D converter.

I will discuss the need for an anti-aliasing filter in the next section. For now, I'll describe the A-D converter, which comprises a sample-and-hold (S/H) function and a quantizer. The sampler is simply a switch which closes every T_S seconds to sample the analog voltage. The switch is closed for a very short time so that the analog signal changes very little, if at all, during the sampling process.

The sample voltage is dumped onto a very high-quality, low-leakage, capacitor which holds the sampled value until the next sample. This gives the quantizer sufficient time to convert the analog voltage into a digital word. The quantizer converts the input voltage range into $2^N - 1$ levels, where N is the number of bits. For a 5V input range, a 12-bit A-D converter will produce a digital output corresponding to 4095 levels, each level being equal to an analog voltage increment of 1.22mV.

6.3.2 SAMPLING THEOREM AND ALIASING

You should have an intuitive sense that you need to sample signals at a rate faster than the signal changes in order to capture an acceptable replica of it. Let's try to make this simple idea more concrete.

6.3.2.1 THE SAMPLING THEOREM

Figures 6.13a–d illustrate the sampling of four sine-wave signals with frequencies of 125, 250, 500, and 1,000Hz. The respective periods of these signals are 8, 4, 2, and 1ms. The signals are sampled at 1ms intervals. Thus, the sample time and sampling frequency are:

$$T_S = 1ms = 0.001s \text{ and } F_S = 1000Hz$$

You should not confuse the sampling frequency, F_S, with the lowercase symbol, f_S, used throughout this book for a driver's resonant frequency.

In *Figs. 6.13* sample values are indicated with a black dot on the plots. Eight samples are produced for every period of the 125Hz sine wave, four samples per period for the 250Hz sine wave, two for 500Hz, and only one for the 1000Hz signal. In general, the sample times do not coincide with zero crossings of the sine wave and—most importantly—*all knowledge of the signal between samples is lost!*

Can you reconstruct the signal from its samples? Looking only at the sample points, and knowing that the sampled signals are sine waves, you might become fairly confident about visually fitting 125Hz and 250Hz sine waves to the sample points in *Figs. 6.13a* and *b*. You would have a great deal more trouble with the 500Hz sine wave in *Fig. 6.13c*, and the constant sample values associated with the 1000Hz sine wave in *Fig. 6.13d* would lead you to believe that a DC voltage was sampled.

A better place to look for signal reconstruction is at the output of the S/H circuit of the A-D converter. The signal at this point is still analog; that is, it is still a continuous-time signal. *Figures 6.14a–d* show the S/H output for the 125, 250, 500, and 1000Hz signals. In each case, the output is a "staircase" approximation of the input.

Examine *Fig. 6.14a*. Note that the stair-case signal is periodic, and, furthermore, its period is the same as that of the input signal, 8ms. From Section 6.2 you know that the stair-case signal has a Fourier series representation. You also know that the fundamental term in the series will be at 125Hz. The stair-case signal contains a replica of the original input! If the S/H output is fed to a sharp cutoff

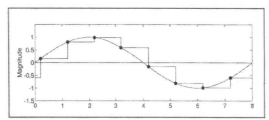

FIGURE 6.14a: Sample-and-hold output at 125Hz.

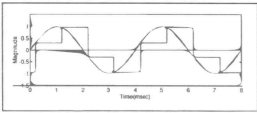

FIGURE 6.14b: Sample-and-hold output at 250Hz.

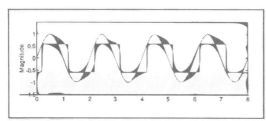

FIGURE 6.14c: Sample-and-hold output at 500Hz.

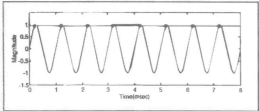

FIGURE 6.14d: Sample-and-hold output at 1000Hz.

low-pass filter with a cutoff frequency somewhat above 125Hz, you can recover the 125Hz signal with not much error, except perhaps for a small phase shift.

Figures 6.14b and *6.14c* show that the output of the S/H is periodic with the same period as the input. In the case of the 500Hz signal, the S/H output is a 500Hz square wave. Again, with the right low-pass filter, you can recover both input signals. The S/H output for the 1000Hz signal is a constant value (*Fig. 6.14d*). You cannot recover a 1000Hz sine wave sampled at a 1000Hz rate.

You can conclude from *Fig. 6.14* that the highest signal frequency which can be recovered when sampling at a 1000Hz rate is 500Hz. This result is generalized by the sampling theorem. Here is one statement of the theorem:[4]

Sampling Theorem: If the highest frequency contained in an analog signal, x(t), is F_{MAX} and the signal is sampled at a rate $F_S \geq 2F_{MAX}$, then the signal can be exactly recovered from its samples with the appropriate interpolation function.

The theorem states what you may have already surmised. If you wish to fully recover a 1000Hz signal from its samples, you must sample it at a 2000Hz rate (at least). Alternatively, at the 1000Hz sample rate, the highest frequency you can recover without error is 500Hz.

The right interpolation function is the $\sin(x)/x$ or sinc function. I will not discuss the sinc function or other more practical reconstruction filters. I am only trying to develop a qualitative understanding of sampling here. Discussion of this arcane aspect of digital signal processing is best left to the references.[3,4] It should be clear, however, that the hold circuit is a very poor reconstruction filter for signals near $F_S/2$.

The sampling frequency $F_N = 2F_{MAX}$ is called the Nyquist rate. The optimum interpolation filter, the sinc function, cannot be realized in practice. Practical reconstruction filters require sampling at a rate somewhat greater than F_N.

6.3.2.2 ALIASING AND THE ANTI-ALIASING FILTER

What happens when you sample a signal at less than the Nyquist frequency? *Figure 6.15* shows the sampling of 250Hz and 750Hz sine waves at a 1000Hz rate. It also shows the output of the S/H block for each signal. Like its 250Hz input, the stair-case output of the 250Hz S/H block has a period of 4ms, but so does the output of the 750Hz S/H. The output fundamental frequency is 250Hz, not 750Hz!

In fact, if you reverse the polarity of the 750Hz S/H output, it is an exact replica of the 250Hz output! This spurious output is called an alias of the original input signal. By under-sampling the 750Hz signal, you have aliased it, as was the case in which a 1000Hz signal sampled at 1000Hz produced a DC output.

The process of sampling a continuous-time signal with a band-limited spectrum produces images of the input spectrum spaced at intervals of the sampling frequency. This leads to a simple equation for predicting the first aliased frequency:

$$F_A = kF_S - F_{in}$$

where:

$$F_A = \text{first aliased frequency}$$

$$F_{in} = \text{analog input frequency}$$

k = smallest positive integer that yields a positive F_A

For the 750Hz signal in *Fig. 6.15*:

$$F_A = 1000 - 750 = 250\text{Hz}$$

In the case of the 1000Hz signal:

$$F_A = 1000 - 1000 = 0\text{Hz or DC}$$

With a 1500Hz signal sampled at 1000Hz:

$$F_A = 2 \times 1000 - 1500 = 500\text{Hz}$$

Now the need for an anti-aliasing filter should be clear. When sampling at 1000Hz, you must be absolutely certain that no frequencies higher than 500Hz are in the sampled signal by placing a sharp cutoff filter with a 500Hz or less cutoff frequency ahead of the S/H block. The required amount of attenuation is a function of the number of bits mechanized in the A-D quantizer. Each bit is worth 6.02dB of dynamic range. For a 12-bit converter, for example, all frequencies higher than $F_S/2$ must be attenuated by at least 12×6.02, or 72.2dB.

Practical analog filters cannot attenuate frequencies arbitrarily fast. The MLSSA PC-based acoustic data-acquisition system described in Chapter 7 uses an eighth-order Chebyshev anti-aliasing filter ahead of a 12-bit A-D converter. This filter produces 72dB attenuation at about 1.5 times its cutoff frequency. To reduce frequencies above $F_S/2$ to acceptable levels with this filter, you must set F_S to three times the filter cutoff frequency. For a 20kHz bandwidth,

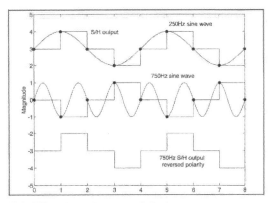

FIGURE 6.15: An illustration of aliasing.

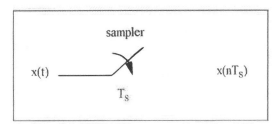

FIGURE 6.16: Converting a continuous-time signal into discrete time samples.

the sample frequency must be 60kHz or more.

6.4 THE DISCRETE FOURIER TRANSFORM

The Discrete Fourier Transform (DFT) and the computationally efficient algorithm for computing a restricted version of it, the Fast Fourier Transform (FFT), are the cornerstones of modern digital signal processing. In this section I will describe the DFT and discuss the concept of resolution, frequency bins, spectral leakage, and data windowing with simple examples to illustrate each point.

6.4.1 THE DISCRETE FOURIER TRANSFORM

Regardless of how long an analog signal may persist, time and/or memory and computing power limitations dictate measuring only a finite length sample of it. The DFT transforms a *finite* length sequence of samples from a periodic time-domain signal into a finite length sequence of harmonically related terms much like a Fourier series. Before I describe the DFT, you must first understand the sampling scenario and some nomenclature. As illustrated in the *Fig. 6.16*, the sampling process converts a continuous-time signal, $x(t)$, into a discrete-time or sampled signal, $x(nT_S)$, where n is an integer.

The period of the sampled signal is assumed to be NT_S seconds, where N is the total number of samples taken. The starting time of the sampling process is arbitrarily set to zero. Thus the N samples span a time period of $(N-1)T_S$ seconds (*Fig. 6.17*). In this figure the sampling process begins at a positive-going zero crossing and proceeds for 25

samples, stopping just before the next positive-going zero crossing.

In digital signal-processing texts, a sampled periodic signal is usually written in terms of a finite series of complex exponentials. I will write it here in terms of cosines to better make the analogy with Fourier series. One form of the DFT representation of a sampled signal, $x(nT_S)$, is written below:

$$x(nT_S) = \frac{1}{N} \sum_{k=0}^{k=\frac{N}{2}} C_k \cos\left[2\pi k f_0 \left(nT_S\right) + \theta_k\right] \quad [6.12]$$

where N is even in the summation above. The coefficients, C_k and θ_k, are the frequency-domain description of the sampled signal, that is, the DFT comprises these terms. Like the Fourier series, the DFT representation of $x(nT_S)$ involves a sum of harmonically related terms. You can make the following analogies with the Fourier series:

$$T = NT_S \text{ and } f_0 = \frac{1}{T} = \frac{1}{NT_S} = \frac{F_S}{N} \quad [6.13]$$

C_k and θ_k are the magnitude and phase of kth harmonic component, and the k=0 term is the DC value of the sampled signal. But here the analogy ends.

Unlike the Fourier series, the DFT signal representation is a *finite* sum of harmonic terms. Furthermore, the DFT coefficients reproduce $x(t)$ only at the sample points, $x(nT_S)$, whereas the Fourier series represented $x(t)$ for every real t. This last point is brought home more forcefully by substituting the expressions for NT_S and f_0 given in Equation 6.13 into the cosine argument

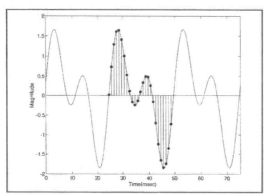

FIGURE 6.17: Periodic signal and samples of one period.

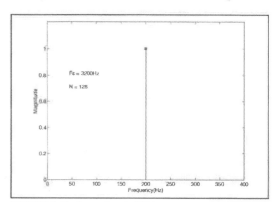

FIGURE 6.19: Example 6.2: FFT magnitude spectrum of 200Hz sine wave.

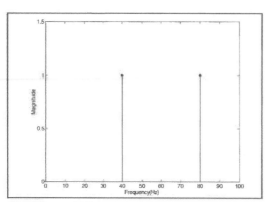

FIGURE 6.18: DFT of periodic signal shown in *Fig. 6.17*.

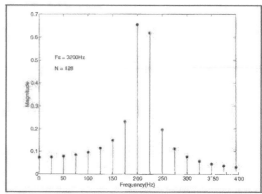

FIGURE 6.20: Example 6.2: FFT magnitude spectrum of 212.5Hz sine wave.

of Equation 6.12:

$$x(nT_s) = \frac{1}{N}\sum_{k=0}^{k=\frac{N}{2}} C_k \cos\left[2\pi k\left(\frac{1}{NT_s}\right)(nT_s) + \theta_k\right] = \frac{1}{N}\sum_{k=0}^{k=N/2} C_k \cos\left[2\pi k\frac{n}{N} + \theta_k\right]$$

[6.14]

The sample time, T_S, disappears, and $x(nT_S)$ is a function of n and k alone, where n is the sample number and k is the order of the harmonic. For this reason it is customary to drop the "T_S," simply referring to $x(nT_S)$ as $x(n)$. For each value of n, Equation 6.14 states that the sample point, $x(n)$, is reproduced by summing the cosine terms in Equation 6.14 over all values of k from 0 to $(N/2)$.

Notice that $x(n)$ is periodic in N, that is, $x(n)$ repeats every N samples because the cosine is periodic. This is true even if the original signal, $x(t)$, is not periodic! This fact will become important later in the discussion.

Finally, as you might well expect, the highest frequency in the sum is $F_S/2 = F_N$, the Nyquist frequency. This is easily shown. The largest value for k is N/2, so that:

$$k \times f_0 \Rightarrow \frac{N}{2} \times \frac{1}{NT_S} = \frac{1}{2T_S} = \frac{F_s}{2} = F_N$$

6.4.1.1 EVALUATING THE COEFFICIENTS C_K AND Θ_K

The modern revolution in digital signal processing can be traced to the publication by Cooley and Tukey in 1965[5] of a highly efficient algorithm for computing a special form of the DFT. This algorithm came to be known as the Fast Fourier Transform (FFT). There is some evidence that the FFT was invented by Carl Friedrich Gauss in 1805, two years before Fourier's first attempt at publishing.[3]

The FFT was known to British radar signal analysts in the late 1930s. They used a version of it implemented on mechanical calculators to analyze German radar signals.[6] Nevertheless, it was the Cooley-Tukey paper that brought the FFT to the general engineering and signal-processing community.

The computational efficiency of the FFT over the DFT comes from taking full advantage of the sine/cosine periodicity (actually, the periodicity of the complex exponential) and limiting the number of samples, N, to the form $N = r^v$, where r, the *radix* of the FFT algorithm, and v are integers. The most common value for r is two, limiting the number of samples, N, to powers of 2.

I will not present the FFT algorithm here. It is covered very well in the references.[3,4] Many software programs are available to compute the FFT from a sequence of samples. It is not necessary to understand the algorithm in detail to use this software effectively. I computed all DFTs and FFTs in this chapter with a program called MAT-LAB™, an engineering software package.

6.4.1.2 EXAMPLE 6.1

The periodic signal shown in *Fig. 6.17* consists of the sum of two sine waves of equal ampli-

tude—one at 40Hz and one at 80Hz. The 80Hz sine wave has a phase angle of $\pi/10$ radians, or 18°. (See Section 6.6 for a discussion of radian

TIME, FREQUENCY, AND THE FOURIER TRANSFORM

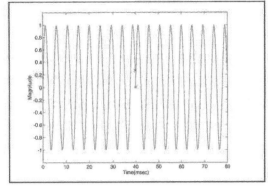

FIGURE 6.21: Concatenation of two sample lengths of 212.5Hz sine wave.

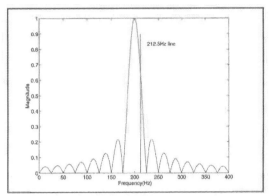

FIGURE 6.22a: Example 6.2: FFT 200Hz bandpass filter response.

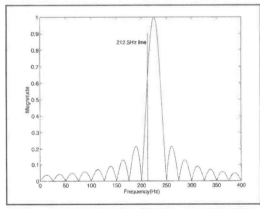

FIGURE 6.22b: Example 6.2: FFT 225Hz bandpass filter response.

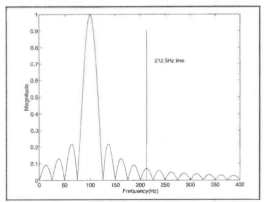

FIGURE 6.22c: Example 6.2: FFT 100Hz bandpass filter response.

101

angular measure.) The equation for the signal is given below:

$$x(t) = \sin(2\pi \times 40t) + \sin(2\pi \times 80t + \pi/10) \quad [6.15]$$

This signal was sampled at a 1kHz rate for 74ms, yielding a total of 75 samples. Use Equation 6.13 to obtain the following:

$$T_S = 1/1000 = 0.001s, \quad f_0 = F_S/N = 1000/75 = 13.3333...Hz$$

The DFT magnitude for this signal (*Fig. 6.18*) consists of two frequency components of equal magnitude at 40 and 80Hz. These frequencies correspond to values of the frequency index, k, equal to 3 and 6.

6.4.2 LEAKAGE AND WINDOWS
The previous example is somewhat artificial. The

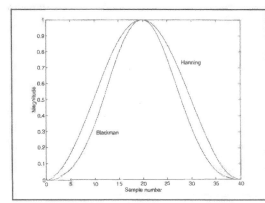

FIGURE 6.23: Hanning and Blackman data windows.

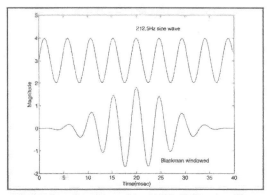

FIGURE 6.24: Original and windowed 212.5Hz sine wave.

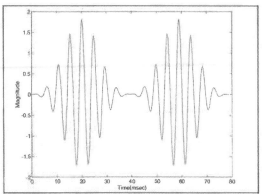

FIGURE 6.25: Two sample lengths of windowed 212.5Hz sine wave.

sample length, NT_S, corresponds exactly to three cycles of the periodic signal so that the signal frequencies fall exactly at integer multiples of f_0. The signal is also periodic over the sample length, since any signal periodic over T seconds is also periodic over integer multiples of T seconds.

In more realistic situations you may not know the signal period before you sample the signal. Even if you do know the period, you may not be able to set the sampling frequency such that the sample length spans an integer number of signal periods. In this case you will experience the phenomenon called spectral leakage.

6.4.2.1 EXAMPLE 6.2: SPECTRAL LEAKAGE
This example examines the FFT of two sine waves. The sampling parameters are as follows:

$$F_S = 3200Hz, N = 2^7 = 128, F_N = 1600Hz, \quad T=N/3200 = 0.04s = 40ms$$

then:

$$f_0 = 3200/128 = 25Hz$$

The two sine waves have frequencies of 200Hz and 212.5Hz. The first sine wave frequency corresponds to the frequency index k = 8. The frequency of the second sine wave falls midway between the k = 8 and k = 9 (200Hz and 225Hz) indices.

The FFT magnitude of the 200Hz sine wave is shown in *Fig. 6.19*. As you would expect, the spectrum consists of a single line at 200Hz. *Figure 6.20* shows the FFT magnitude of the 212.5Hz sine wave. The frequency scale in both figures is linear and has been expanded to show only the first 400Hz of the spectrum (up to k = 17). The 212.5Hz sine-wave spectrum shows components of almost equal magnitude at 200 and 225Hz with additional components both above and below those frequencies with magnitudes that decrease with increasing distance from the 200 and 225Hz points. This is *spectral leakage*.

With the sampling parameters set up in this example, the FFT can only resolve frequencies in increments of 25Hz. The 212.5Hz sine wave falls between two 25Hz positions. The best the FFT algorithm can do is to assign signal energy to the two nearest frequencies. Unfortunately, additional signal energy "leaks" into many of the neighboring positions, hence the terminology "leakage."

Here is a qualitative explanation of spectral leakage. In the above example, the FFT must model *all* frequencies in the signal with a finite set of harmonic terms spaced 25Hz apart. Each term in the FFT completes a whole number of cycles over the sample period of 40ms. The 25Hz fundamental goes through one complete cycle.

The second harmonic completes two cycles while the kth harmonic completes k cycles. If a particular harmonic starts the 40ms with a particular value and slope, it will end the 40ms interval with the same value and slope. Each term in the FFT series is periodic over the sample interval.

Remember, the FFT model assumes the signal is infinite in length, but periodic over the 40ms sampling interval. If there is a frequency in the signal which is not periodic over 40ms, its value at the end of the 40ms interval will not be the same as its value at the beginning of the period. This is the case for the 212.5Hz tone.

Two 40ms periods of the 212.5Hz signal that the FFT is trying to model are shown in *Fig. 6.21*. The end of the first sample period and the beginning of the second sample period are both marked with an "x." The 212.5Hz signal is passing through zero with a positive slope at the beginning of the first sample interval, but it has a positive, non-zero value at the end of that interval with a negative slope, i.e., the signal is decreasing in value.

At the beginning of the next interval, the signal being modeled by the FFT is again zero. Unlike the original 212.5Hz sine wave, which is continuous for all time, the periodic signal modeled by the FFT has an instantaneous jump from the end of one interval to the start of the next. This jump introduces spurious frequencies which show up as leakage.

Note that concatenating two 40ms sample intervals of the 200Hz sine wave would not have this jump because the 200Hz sine wave will fit evenly in any number of contiguous 40ms periods with the same value at the beginning and end of each 40ms interval.

6.4.2.2 THE FFT AS A BANK OF BANDPASS FILTERS

For a more technical explanation of leakage, consider the FFT as a bank of bandpass filters. You saw one such bank of filters before when I discussed the one-third octave RTA in Chapter 4. The FFT acts like a bank of bandpass filters with center frequencies spaced at f_0 Hz apart.

In Example 6.2, the filters are spaced every 25Hz. The FFT bandpass filters, however, behave very differently from the one-third octave analog filters in an RTA. The 200Hz, 225Hz, and 100Hz bandpass filters for Example 6.2 are shown in *Figs. 6.22a–6.22c*, respectively.

Look at the 200Hz filter response in *Fig. 6.22a*. Response is one at 200Hz where k = 8 and zero for every other value of k. For example, the 200Hz bandpass filter will not respond at all to frequencies of 150, 175, 225, or 250Hz. It will, however, respond to frequencies in the sampled signal that fall between the 25Hz intervals.

In Example 6.2, the 200Hz and 225Hz filters respond almost equally to the 212.5Hz sine wave, producing the two largest lines in *Fig. 6.20*. The 100Hz filter also responds to the 212.5Hz sine wave. It is the bandpass filter non-zero response to frequencies between multiples of f_0 that causes leakage.

From *Fig. 6.22* it is clear that the FFT in Example 6.2 cannot resolve multiple signals with frequencies closer together than 25Hz. For example, sine waves of 190, 200, and 210Hz will all appear with the greatest magnitude at the 200Hz spectral line. If all three signals are present they will appear as one large line at 200Hz, with smaller leakage components above and below 200Hz. In general, the FFT cannot resolve frequencies spaced closer than f_0. This limit on resolution leads to the concept of frequency "bins."

A frequency bin is simply an alternate way of expressing the frequency-resolution capability of the FFT. The bin defines an interval in frequency. It has width f_0 centered on the center frequency of an FFT bandpass filter. The 200Hz bin in Example 6.2 is 25Hz wide and extends from 187.5Hz to 212.5Hz. In any spectral analysis of a signal, all frequencies within this interval will be assigned to the 200Hz spectral line.

Of course, frequency components of the signal not exactly at 200Hz will also leak into other bins, but they will show up with greatest magnitude at the 200Hz line. More generally, the FFT acts like a sorting process, assigning all frequencies in the range $kf_0 \pm f_0/2$ to the kth spectral line.

6.4.2.3 REDUCING LEAKAGE WITH DATA WINDOWS

Leakage occurs when the signal being sampled is

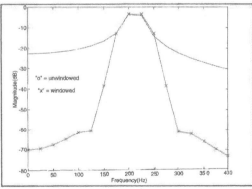

FIGURE 6.26: Example 6.2: Spectrum of windowed 212.5Hz sine wave.

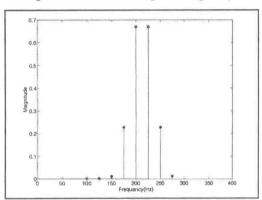

FIGURE 6.27: Example 6.2: Windowed versus unwindowed 212.5Hz FFT spectra.

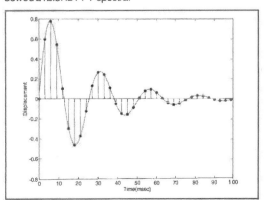

FIGURE 6.28: Spring-mass impulse response and its samples.

not periodic in the sample interval, T. We can force the signal to be periodic in T by multiplying it by a *data window* that smoothly tapers the signal to zero at each end of the sample period. There are many data windows, each with unique benefits.[7] Two commonly used ones are the Hanning and Blackman windows. The equations describing these windows are given below:

Hanning window:

$$w_H(n) = 0.5[1 - \cos(2\pi n/N)], \; n = 0, 1, 2, 3...N - 1$$

Blackman window:

$$w_B(n) = 0.42 - 0.5\cos(2\pi n/N) + 0.08\cos(4\pi n/N),$$
$$n = 0, 1, 2, 3...N - 1$$

These windows are graphed in *Fig. 6.23*. They appear quite similar, but the Blackman window

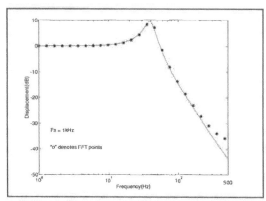

FIGURE 6.29: Comparison of FFT and FT of spring-mass frequency response.

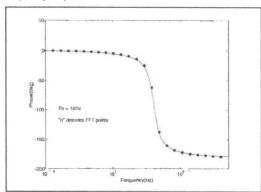

FIGURE 6.30: Comparison of FFT and FT of spring-mass phase response.

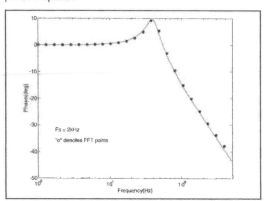

FIGURE 6.31: Comparison of FFT and FT of spring-mass phase response.

rises more slowly from either end toward the center of the data interval. Thus the Blackman window concentrates more heavily on the data in the middle of the sample interval. Note that the power in a signal is reduced when it is multiplied by a non-rectangular data window.

Continuing with Example 6.2, *Fig. 6.24* shows the original 212.5Hz sine wave and the result after multiplying the 128 samples of this signal point-by-point by a 128-point Blackman window. The windowed signal has been rescaled to produce the same *power* as the original sine wave. The windowed signal now resembles a shaped tone burst with zero values at each end of the sample period.

If you concatenate two of the windowed signals together, you get the result shown in *Fig 6.25*. Unlike the unwindowed signal, there is now no jump discontinuity at 40ms. The windowed sample can now be viewed as one period of an infinitely long data stream that is periodic over 40ms.

The FFT of the windowed signal is shown in *Fig. 6.26*. Compare this result against *Fig. 6.20*. There are still two equal components at 200 and 225Hz. You cannot overcome the basic resolution capability of the FFT. But all of the leakage components below 175Hz and above 250Hz have been greatly reduced.

Figure 6.27 shows a better view of the windowing effect. Here the FFT magnitude is plotted on a dB scale. Unwindowed FFT points are marked with an "o" and windowed FFT points with an "x." Leakage with the unwindowed signal is only 20–25dB below the major lines.

At 150Hz and 275Hz, leakage from the windowed signal is 23dB below the unwindowed leakage components. This number improves to 50dB or more below 125Hz and above 300Hz. Windowing does decrease resolution somewhat depending on the window used, but this is a small price to pay for the reduced leakage.

6.5 THE FFT OF THE IMPULSE RESPONSE

This section examines the FFT of the spring-mass impulse response shown in *Fig. 6.3* and compares it to its frequency response obtained from the continuous-frequency FT. It also covers impulse-response windowing and zero padding of the data sample.

6.5.1 EXAMPLE 6.3: FFT OF THE SPRING-MASS SYSTEM IMPULSE RESPONSE

The spring-mass system displacement frequency response shown in *Fig. 6.3* looks like the response of a slightly under-damped low-pass filter with a 40Hz cutoff frequency. This low-pass response may reduce or possibly eliminate the need for an anti-aliasing filter as long as the sampling frequency is high enough.

Let's see what happens with a 1kHz sampling frequency. The first 100ms of the sampled version of the spring-mass displacement impulse response is shown in *Fig. 6.28*. To avoid excessive clutter on the plot, only every third sample point is shown. The sampling parameters are given below:

$$F_S = 1000Hz, N = 2^{10} = 1024, F_N = 500Hz,$$
$$T = N/1000 = 1.024s$$

then:

$$f_0 = 1000/1024 = 0.9766Hz$$

The magnitude of the 1024-point FFT is plotted against the true continuous-frequency frequency response in *Fig. 6.29*. Again, to avoid excessive clutter on the plot only a subset of the actual number of FFT points is shown. The FFT looks like a sampled version of the FT. The sample points track the true frequency response well out to about 150Hz. Beyond that frequency the FFT values rise above the true response due to aliasing. The 1024-point FFT-derived frequency response is 8dB above the true response at 500Hz.

The phase response of the 1024-point FFT is plotted against the true phase response in *Fig. 6.30*. The FFT phase response tracks the true phase response with little error. For this example, phase response is less sensitive to aliasing than magnitude response.

Increasing the sampling frequency to 2kHz produces the following sampling parameters:

$$F_S = 2000Hz, N = 2^{11} = 2048, F_N = 1000Hz, T = N/$$
$$2000 = 1.024s$$

then:

$$f_0 = 2000/2048 = 0.9766Hz$$

Notice I have doubled the sampling frequency without increasing the sample interval, T. Thus, f_0 is unchanged, but F_N is doubled. The magnitude of the 2048-point FFT is plotted against the true frequency response in *Fig. 6.31*. The 2048-point FFT now tracks the true frequency response well out to about 300Hz. Although greatly reduced, there is still some aliasing of the 2048-point FFT-derived frequency response above 300Hz. At 500Hz the FFT-derived response is roughly 3dB above the true response.

The original continuous-frequency magnitude response and its aliases are shown in *Fig. 6.32*. The 2kHz alias has 12dB less power at 500Hz than the 1kHz alias, which explains the reduction in frequency-response error. A further increase of the sampling frequency to 4kHz will produce a frequency response that is accurate out to 500Hz. Using an eighth-order Chebyshev anti-aliasing filter with a 1kHz cut-off frequency together with a 4kHz sampling rate will produce a frequency response accurate to 1kHz. (This requires additional post-processing to remove the known effect of the anti-aliasing filter.)

Based on the example in this section, you can make the following generalization:

Under the *right* conditions the FFT of the time-sampled impulse response is a frequency-sampled version of the continuous-frequency frequency response.

The right conditions are:

1. The sampling frequency is high enough and/or the proper anti-aliasing filter is employed to prevent aliasing out to the highest frequency of interest.
2. The impulse response has decayed essentially to zero over the sample interval, $NT_S = T$.

The first condition is obvious. The second condition guarantees that the impulse response is zero at both ends of the sample period so that it may be considered as one period of a signal that is periodic over T seconds. A periodic sequence of impulse responses was shown in *Fig. 6.6*. Only one full period is needed to compute the FFT.

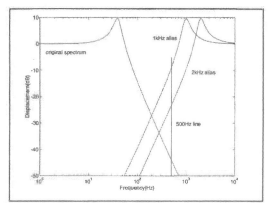

FIGURE 6.32: Original and aliased spectra.

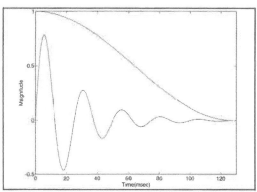

FIGURE 6.33: Impulse response and 256-point half Hanning window.

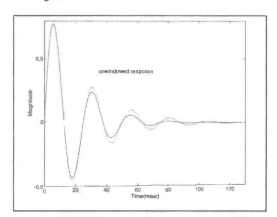

FIGURE 6.34: Windowed versus unwindowed impulse response.

6.5.2 EXAMPLE 6.4: WINDOWING THE IMPULSE RESPONSE

If the impulse response has not decayed to zero at the end of the sample interval, you can use a window to force the zero value condition. By definition, the impulse response is zero at time zero. You

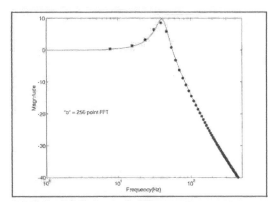

FIGURE 6.35: 256-point FFT versus 2048-point FFT.

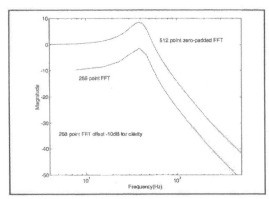

FIGURE 6.36: Effect of zero-padding on FFT.

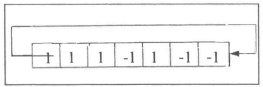

FIGURE 6.37a: Length seven MLS sequence.

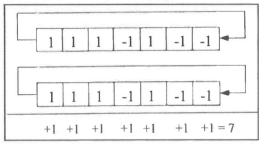

FIGURE 6.37b: Autocorrelation of MLS sequence: Shift = 0.

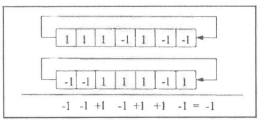

FIGURE 6.37c: Autocorrelation of MLS sequence: shift = 5.

need only enforce the zero condition at the end of the sample period. The proper window for windowing an impulse response is the *half-window*.

Suppose you have available only the first 256 samples of the spring-mass impulse response sampled at a 2kHz rate. The sampling parameters become:

$$F_S = 2000Hz, N = 2^8 = 256, F_N = 1000Hz,$$
$$T = 256/2000 = 0.128s$$

then:

$$f_0 = 2000/256 = 7.8125Hz$$

The first 128ms of the impulse response and a Hanning half window are shown in *Fig. 6.33*. The value of the impulse response at 1.024s is -8.85×10^{-20}, which is 379dB below its peak value of 0.78. This level is essentially zero, fulfilling condition 2 of Section 6.5.1. At the end of 256 samples, or 128ms, however, the value is 0.0028, which is only 49dB below its peak value and well within the resolution capability of a 12-bit A-D converter. Clearly, you must window the shortened impulse response before taking its FFT.

Windowed and unwindowed versions of the impulse response are shown in *Fig. 6.34*. Ripples in the unwindowed response are clearly evident out to 120ms. These ripples are suppressed by the Hanning window. The 256-point FFT is compared against the 2048-point FFT of Example 6.3 in *Fig. 6.35*. You have lost a factor of 8 in resolution, (f_0 is eight times larger), but otherwise the two FFTs agree reasonably well. The windowed response peaks about 2dB less due to some loss of energy in that region caused by the window, and the peak is somewhat broader, but the similarities are more impressive than the differences, given the factor of 8 reduction in data. If the shortened impulse response had not been windowed, significant leakage would have occurred.

6.5.3 EXAMPLE 6.5: ZERO-PADDING

One final example will end this section on the FFT. The FFT can be used as its own interpolator. In the previous example, the 256-point FFT resulted in frequency points spaced every 7.8125Hz. This spacing produces a rather coarse frequency-response plot. You can get points in between by the process of zero-padding.

In zero-padding, extra points are added to the end of the data sample. All of the added points are zero in value. The signal is not changed, but the sample length is increased with the fictitious points. If you add 256 zero-valued points to the shortened sample in Example 6.4, you get a new data sample of length 512. The sampling parameters for this sample length are:

$$F_S = 2000Hz, N = 2^9 = 512, F_N = 1000Hz,$$
$$T = 512/2000 = 0.256s$$

then:

$$f_0 = 2000/512 = 3.90625Hz$$

The artifact of zero padding has increased the number of frequency points by 2 and cut the frequency spacing in half. The resulting 512-point FFT magnitude is compared against the 256-point curve in *Fig. 6.36*. The original 256 points are unchanged. Zero-padding has added an additional 256 points, one each between the original points. The additional points make the curve much smoother, especially in the region around the response peak at 40Hz. Zero-padding is routinely used in the CLIO and MLSSA PC-based systems to be discussed in Chapter 7.

You may think that you got something for nothing with zero padding. This is not so. The extra points are interpolated. There is no new data and the resolution is unchanged. The response peak just below 40Hz is still 2dB less and somewhat broader than it should be. All you have obtained is a somewhat smoother curve.

6.6 A PRACTICAL WAY TO MEASURE THE IMPULSE RESPONSE

As discussed in Section 6.2, some earlier PC-based acoustic data-acquisition and analysis systems used a narrow rectangular pulse to approximate an impulse. Pulse widths under 10μs were typical. This width corresponds to less than 20% of the period of a 20kHz sine wave. The pulse stimulus is very brief, but data must be collected over a period of up to tens of milliseconds to obtain good low-frequency information. During this extended period, the test microphone will pick up not only the DUT decay response, but also any noise present in the test environment.

These narrow pulses contain little energy. When this small amount of energy is spread over the entire audio spectrum, even less energy is presented to each FFT frequency bin. The result is a DUT response with very poor signal-to-noise ratio (SNR). Increasing the height of the pulse will put more energy into the process, but very high pulse amplitudes can lead to nonlinear behavior of the DUT. SNR can be improved by running several test trials and averaging the results to reduce the noise.

White noise is a signal with a flat power spectrum. It might seem like an ideal test signal since its crest factor, the ratio of peak value to RMS value, is relatively low. Instead of putting all the energy into the DUT within 10μs or less, the test signal can be applied much longer. With the proper signal-processing algorithm, you can obtain very good SNRs without driving the DUT into nonlinear operation.

There are some problems with white noise, however. First of all, it is a completely random signal. Since the signal is unknown before the test, both the input white noise and the resulting system response must be recorded, requiring a dual-channel analyzer. Again, because it is completely random, its power spectrum is flat only when averaged over a significant time period. From the standpoint of the FFT, the white noise is not periodic. Strictly speaking, the FFT of white noise does not exist. (The spectrum of white noise refers to its *power* spectrum. Its *signal* spectrum

will change from sample period to sample period.)

Is there another signal with all the desirable properties of white noise and few of its drawbacks? Yes. It is the maximum length sequence (MLS) signal. An MLS is a discrete number sequence that switches between two values in an almost random fashion. For this reason it is also called a pseudo-random noise (PN) sequence.

The sequence can be generated with a shift register and one or more Exclusive OR (XOR) gates. A feedback loop is established by tapping the output of two or more stages and feeding the result back to the first stage. If the shift register has N stages, the length, L, of the longest sequence that can be obtained before the sequence starts to repeat is L = $2^N - 1$. If you choose the taps and the XORing properly, the resulting binary sequence will be maximum length.

For our purposes, an MLS signal has two very important features: 1) even though it looks ran-

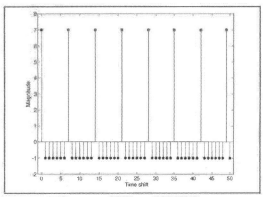

FIGURE 6.38: Autocorrelation function of length seven MLS signal.

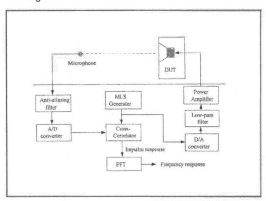

FIGURE 6.39: Block diagram of MLS impulse/frequency response measurement.

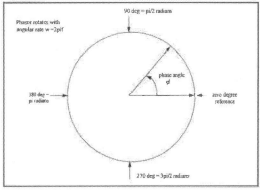

FIGURE 6.40: Sine-wave phasor.

dom, it is not; it is deterministic and exactly repeatable; and 2) it is periodic with a period length of $2^N - 1$ times the shift register clock rate. The first feature means you do not need a dual channel analyzer since the input signal is known. The second means that the output signal is also periodic and you can compute FFTs without leakage if the entire data period is used.

The output of the shift register is usually level-shifted so the MLS switches between values of +1 and −1. Since there is always one more +1 than −1 value, the MLS has a DC value equal to $1/L$. The MLS power spectrum equals $(L + 1)/L$ at all other discrete frequencies spaced F_c/L apart, where F_c is

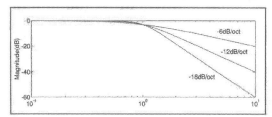

FIGURE 6.41a: Low-pass filter frequency responses.

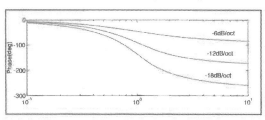

FIGURE 6.41b: Low-pass filter phase responses.

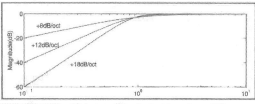

FIGURE 6.42a: High-pass filter frequency responses.

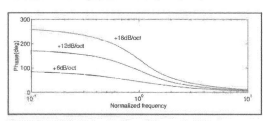

FIGURE 6.42b: High-pass filter phase responses.

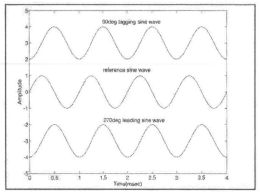

FIGURE 6.43: Oscilloscope traces showing phase measurement ambiguity.

the shift register clocking frequency.

Thus, the MLS power spectrum looks like discrete-time white noise.[9,10]

As an example, consider a 15-bit shift register clocked at 60kHz. The period of the MLS, T_{MLS}, signal will be:

$$T_{MLS} = L/F_c = (2^{15} - 1)/60,000 = 32,767/60,000 = 0.5461s$$

The fundamental frequency is then:

$$f_0 = 1/T_{MLS} = F_c/L = 1.83Hz$$

The spectrum DC value is $1/32767$, or 3.05×10^{-5}. At all other discrete frequencies the spectrum amplitude will equal $32768/32767 = 1.00003$, which is unity for all practical purposes. The spectral lines will be spaced 1.83Hz apart. In one complete period, each discrete frequency will contain the same power.

So far, this discussion has treated the MLS signal as an abstract sequence of +1 and −1 values. The output of the shift register generating the MLS signal is actually a series of rectangular pulses with a width of one clock period. The spectrum is now no longer flat. It has a sinc squared $[\sin(\omega)/\omega]^2$ shape which is down 1dB at one-third the clock frequency. This does not present a problem in practice, since the known spectrum rolloff can be compensated out in the signal processing. An example of this compensation will be given in Chapter 7.

Summarizing the discussion so far, two important properties of the MLS signal make it ideal for determining the impulse response of a system: 1) like discrete white noise, its autocorrelation function is an impulse; and 2) the cross-correlation of a system response to an MLS signal with the MLS signal itself is that system's impulse response.[10] The meaning and importance of these properties can be clarified by the following example, which follows along the lines presented in Reference 11.

6.6.1 EXAMPLE 6.6: AUTOCORRELATION AND CROSS-CORRELATION WITH THE MLS SIGNAL

Autocorrelation: A length seven MLS is shown in *Fig. 6.37a*. After the last value, the sequence returns to the beginning and repeats. If a second copy of this sequence is lined up below the first and the corresponding values are multiplied together and their products summed, the resulting value is seven. This is shown in *Fig. 6.37b*.

Because the MLS signal is periodic, a circular shift of digits to the left corresponds to a time delay. Any digits that run off the left circulate back onto the right. If the same multiply-and-sum operation is performed with a time-shifted version of the sequence in the lower position, the result is minus 1. This is true for *any* non-zero value of the time shift. A five-place shift is shown in *Fig. 6.37c*.

This process of time shifting, multiplying, and adding a signal with itself is called autocorrelation. *Figure 6.38* is a plot of the autocorrelation function for this example versus time shift. The plot shows a large peak at zero shift and multiples of seven and small negative values elsewhere. This plot is quite similar to the periodic impulse shown in *Fig. 6.6*.

For N = 15, the value of the peaks would be 32,767, and they would be spaced every 32,767 time shifts. For the intermediate time shifts, however, the value of the autocorrelation function is still only –1! It is customary to normalize the MLS autocorrelation function by dividing the summed products by $L = 2^N - 1$. With this normalization the peaks are one and the intermediate shift values are equal to –1/32,767, which is -3.05×10^{-5}. For large N, the periodic MLS signal effectively correlates with itself only at zero time shift and times shifts that are integer multiples of L. But this is exactly the property of discrete white noise.

Cross-correlation: The binary MLS sequence at the output of the shift register looks something like a square wave with random frequency. Suppose now that you feed this signal into the system to be tested and further suppose that you take a sampled version of the resulting output and perform the multiply-and-sum operation of that sample against the MLS input for every circular shift.

The result of this operation, called *cross-correlation*, will be the system's impulse response. The benefit from the MLS signal and the more complex signal processing associated with the cross-correlation operation is that the resulting impulse response has the noise immunity equivalent to averaging the impulse responses from 2^N pulses. For N = 15, this corresponds to about 45dB of noise reduction! This is a phenomenal processing gain.

A block diagram showing the signal processing required to obtain an impulse response with the MLS signal is presented in *Fig. 6.39*, which is adapted from Reference 8. The output of the signal-processing chain is the DUT impulse response and frequency response, both magnitude and phase. Several commercial PC-based acoustic data-acquisition and analysis systems use an MLS stimulus. A partial, but not exhaustive, list includes DRA's MLSSA, Audiomatica's CLIO, Liberty AudioSuite's IMP, the Ariel SYSid, and Audio Precision's System Two.

6.7 TIME VERSUS PHASE

In this section I will define phase precisely and take a look at phase response, minimum phase systems, and excess phase. I will also show the relationship between phase and time delay, and explain the use of group delay.

6.7.1 PHASE AND SINE WAVES

Phase is really all about sine waves. The phase of a sine wave is simply its argument. For example, in the expression

$$x(t) = \sin(2\pi ft + \varphi_0)$$

the bracketed term $(2\pi ft + \varphi_0)$ is the *total phase* of the sine wave, where f is frequency and φ_0 is the initial phase at time zero, the start of the sampling or measurement interval.

Why the 2π? The angular measure used to define the sine, and indeed all trigonometric functions, is

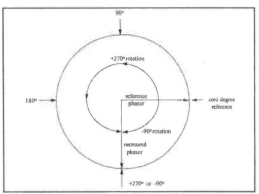

FIGURE 6.44: Alternate view of phase measurement problem.

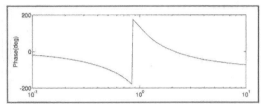

FIGURE 6.45a: Measured phase response of fifth-order low-pass filter.

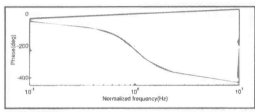

FIGURE 6.45b: Unwrapped phase response of fifth-order low-pass filter.

the radian. When using a handheld calculator to compute the sine, you are probably accustomed to entering a value in degrees, but the calculator converts this value to radians before entering its algorithm for computing the sine.

An angle measured in radians is simply the distance traveled along the circumference of a circle of radius r, divided by r. Thus, one complete trip around the circumference is equal to 2π radians, or 360°. So

$$2\pi \text{ radians} = 6.2832 \text{ radians} = 360°$$

and:

$$1 \text{ radian} = 360/6.2832 = 57.2958°$$

A sine wave goes through one complete cycle in 360° or 2π radians. Ignoring any initial phase and recognizing that $f = 1/T$, you can write the sine argument in the following way:

$$2\pi ft = 2\pi\left(\frac{t}{T}\right)$$

The bracketed expression, (t/T), is the number of whole cycles plus fractions of a cycle through which the sine wave has progressed. Because the sine is periodic, only the fractional part of the argument is needed to compute the sine, but the total phase angle continues to build indefinitely. Some loudspeaker systems have total phase shifts of a few hun-

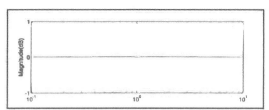

FIGURE 6.46a: All-pass filter frequency response.

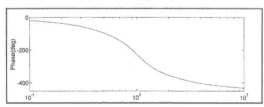

FIGURE 6.46b: All-pass filter phase response.

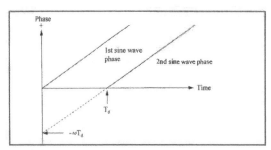

FIGURE 6.47: Phase growth of primary and delayed sine waves.

dred degrees at 20kHz.

f is a circular frequency measure. It is expressed in units of cycles per second, or Hertz (Hz). Engineers and mathematicians prefer to use an angular frequency measure, namely, radians per second. Since one cycle equals 360°, which equals 2π radians:

$$2\pi \text{ radians/s} = 1 \text{ cycle/s} = 1\text{Hz}$$

The Greek symbol, ω, is commonly used for radian frequency, where $\omega = 2\pi f$. Using radian frequency, the sine wave can be written as follows:

$$\sin(2\pi ft + \varphi_0) = \sin(\omega t + \varphi_o)$$

You can see one advantage to using ω; it saves writing "$2\pi f$" over and over again. This symbol, ω, will come up again when I discuss the relationship between time delay and phase in Section 6.7.3.

The amplitude and phase of a sine wave can be represented by a rotating phasor as shown in *Fig. 6.40*. The phasor rotates around the circle at an angular rate f cycles per second, or ω radians per second. By convention, a counterclockwise rotation represents a positive phase growth.

The length of the phasor is the sine-wave peak magnitude, and its position on the circle represents its fractional phase. The phasor may make several revolutions with time, but the total revolution count is lost in this representation. Phasors are helpful in visualizing how two sine waves with the same frequency but different phase will add together.

6.7.2 PHASE RESPONSE AND PHASE MEASUREMENT

The phase of any frequency magnitude response curve is a function of the slope of that curve. This property is illustrated in *Figs. 6.41* and *6.42*. *Figure 6.41a* shows the magnitude response of low-pass Butterworth filters of order 1, 2, and 3 with slopes of −6, −12, and −18dB/octave above their cutoff frequency. The phase response of the three filters is shown in *Fig. 6.41b*.

Phase response is simply the phase angle of the output sine wave *relative* to the input sine wave as a function of frequency. At very low frequencies, the magnitude response slopes are flat and phase shift is very near zero. As frequency increases, the magnitude slopes grow increasingly *negative* and the phase angles also grow more negative, reaching asymptotic values of −90°, −180°, and −270°, respectively, at very high frequencies. Each filter order adds an additional 90° of asymptotic phase. Thus, fifth- and higher-order filters have asymptotic phase shifts greater than 360°.

Figure 6.42 shows magnitude response curves for three high-pass filters with *positive* slopes of +6, +12, and +18dB/octave. The behavior here is just the opposite of the low-pass filters. The rising positive slopes produce positive phase angles. At very low frequencies, the phase angle has asymptotic values of +90°, +180°, and +270°. Each filter order produces an additional +90° phase shift at very low frequencies. At higher frequencies both the magnitude response slopes and the positive phase angles decrease, ultimately becoming zero when the magnitude response flattens out at 0dB.

If the output sine-wave phase is negative relative to the input sine-wave phase, the output is said to *lag* the input. If the output phase angle is positive, the output *leads* the input. The phase periodicity produces an ambiguity in phase measurements, however, that prevents measuring phase angles greater than ±180°.

Figure 6.43 illustrates the problem, showing a three-channel oscilloscope displaying three 1kHz sine waves. The oscilloscope sweep is set to trigger on the positive-going zero crossing of the reference sine wave shown in the middle trace. This sets the start of the time axis for the scope display. The upper scope trace is that of a second 1kHz sine wave with a −90° phase lag. A sine wave leading the reference by +270° is shown in the lower trace. The upper and lower traces are seen to be identical. Looking at solely these traces, you cannot tell whether the phase angle is −90° or +270°.

The phasor diagram of *Fig. 6.44* presents an alternate view of the phase-measurement problem. In this figure the input sine wave is taken as the reference and its phase angle is set to zero. This is equivalent to setting the oscilloscope sweep trigger point. An output phasor is also shown. The question is: How did it get to that position?

There are two possible answers. It could have rotated clockwise through an angle of −90° or counter-clockwise through an angle of +270°. That is, it either leads the input by 270° or lags it by 90°. By convention, the smaller angle is chosen as the correct answer when making phase measurements, so you would say the phase is 90° lagging in this case.

If phase measurements are limited to ±180°, how are phase shifts greater than ±180° determined? The measured phase response of a fifth-order low-pass filter is plotted in *Fig. 6.45a* as a function of frequency. Phase is negative or lagging at low frequencies, as you would expect of a low-pass filter.

Measured phase continues to fall until approximately 0.85 times the cutoff frequency, where it suddenly jumps up from ~180° to +180°. This is the point where the output signal phasor passes the –180° point and is now taken to be a positive value. Beyond that point measured phase again falls with frequency, passing through zero phase and finally reaching –90° well above cutoff frequency.

The jump in measured phase response is an artifact of the measurement process. You can remove this positive jump of 360° by subtracting 360° from all measured phase values after the jump. This produces the true phase response shown in *Fig. 6.45b*. The process of removing the measured phase response jumps is called *phase unwrapping*. *Figure 6.45a* displays the "wrapped" phase while *Fig. 6.45b* shows the "unwrapped" phase. In general, unwrapping measured phase is more complicated than this simple example. Additional examples of unwrapped phase are given in Chapter 7.

6.7.3 MINIMUM PHASE SYSTEMS
In a *minimum phase* system, there is no more phase shift than that dictated by the slope of the magnitude response curve. If the slope is positive, phase will increase in that frequency region. If the slope is negative, phase will decrease. If the slope is flat, phase will remain constant.

If a device is minimum phase, then its phase response is directly related to its magnitude response. Mathematically, the phase response of a minimum phase system equals the Hilbert transform of its log magnitude response (i.e., its magnitude expressed in dB). Any phase response greater than that dictated by the magnitude response is called *excess phase*.

The responses shown in *Figs. 6.41* and *6.42* are minimum phase. Drivers tend to be minimum phase also, at least over their useful frequency range. As discussed in Chapter 4, this property is used by quite a few crossover optimization programs to compute a driver's phase response without directly measuring it. In Section 6.8, you'll see how this minimum phase characteristic helps determine a driver's acoustic phase center.

Fig. 6.46 shows a nonminimum-phase response. This is the response of a second-order Bessel *all-pass* filter. It is called an all-pass filter because the magnitude response is one (0dB) at all frequencies so that it passes all frequencies without attenuation or amplification. A minimum-phase system with this magnitude response has zero phase shift at all frequencies. It is the magnitude response of a straight wire.

But the all-pass network has substantial phase shift as *Fig. 6.46b* clearly shows. All of the phase response of an all-pass network is *excess* phase. All-pass filters are often used as phase equalizers. They can also approximate a pure time delay at frequencies that are low relative to their characteristic frequency.

6.7.4 TIME DELAY AND LINEAR PHASE
Consider this experiment. A sine-wave oscillator is turned on. The turn-on time is taken as the reference time and arbitrarily set to zero. T_d seconds later an identical oscillator is turned on (I am ignoring turn-on transients). What does the phase difference between the two sine waves look like as a function of time?

The phase for both oscillators is plotted in *Fig. 6.47*. They both grow linearly with time at the same rate of ω rad/sec, but the phase of the second oscillator lags the first for all time. The phase difference or phase lag, ϕ_{12}, is obtained by extending the second line back to zero time. From the figure, you can determine:

$$\phi_{12} = -T_d\omega = -(2\pi T_d)f \qquad [6.16]$$

This is true for all frequencies. The magnitude response for a pure time delay is one or 0dB. Phase response for a delay of 0.1ms is shown in *Fig. 6.48a* plotted on the usual logarithmic frequency scale. This plot is not too informative. When plotted on a

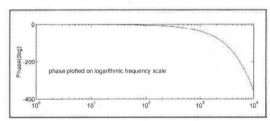

FIGURE 6.48a: Phase response of 0.1ms time delay.

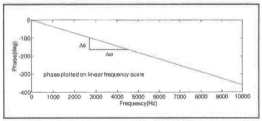

FIGURE 6.48b: Phase response of 0.1ms time delay.

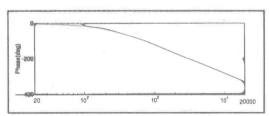

FIGURE 6.49a: Phase response of summed fourth-order L-R crossover outputs.

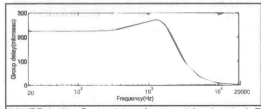

FIGURE 6.49b: Group delay of summed fourth-order L-R crossover outputs.

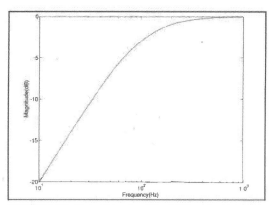

FIGURE 6.50: Frequency response of first-order high-pass filter.

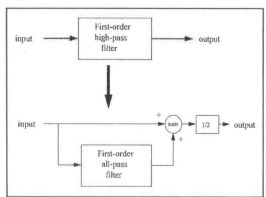

FIGURE 6.51: First-order high-pass filter and its straight-wire/all-pass realization.

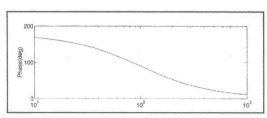

FIGURE 6.52a: Phase response of first-order all-pass network.

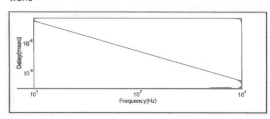

FIGURE 6.52b: Group delay of first-order all-pass network.

linear frequency scale in *Fig. 6.48b*, however, the phase response is a straight line that decreases *linearly* with frequency; the phase response of a pure time delay is a straight line with a negative slope. This is what is meant by *linear phase*.

If radian frequency is used, the phase slope is $-T_d$. With circular frequency the slope is $-2\pi T_d$. The value in seconds is the same in either case.

From *Fig. 6.48b* you can see that a delay of 0.1ms leads to a phase lag of 360° at 10kHz, as you would expect, since the period of a 10kHz sine wave is precisely 0.1ms. A pure time delay does not distort the signal. If it did, you could never play records and CDs. It merely delays the signal to be heard at a later time.

6.7.5 GROUP DELAY

Group delay is a rather difficult quantity to explain. I hope this development will give you some understanding of the concept. The slope of the time-delay phase plot is defined as the incremental change in phase, $\Delta\phi$, for an incremental change in radian frequency, $\Delta\omega$. From *Fig. 6.48b* the negative of the slope for the pure time delay is:

$$\text{slope} = \frac{\Delta\phi}{\Delta\omega} = T_d \qquad [6.17]$$

In this equation $\Delta\phi$ is in radians and $\Delta\omega$ is in radians/second. Thus the slope has units of time. By convention, a negative time delay is quoted as a positive number. If you allow $\Delta\omega$ to become very small, in the limit you'll get an expression for *group delay*, G_d, which is the negative of the derivative of the phase curve with respect to radian frequency:

$$G_d = -\frac{d\phi}{d\omega} \qquad [6.18]$$

This expression also has units of time. Group delay is a local expression, however. It tells how well time relationships between a small group of frequencies are preserved in a narrow frequency region around the point in frequency where G_d is evaluated. You must be very careful when interpreting group delay. As Heyser[11] pointed out, group delay is not equivalent to time delay in general.

There are two important cases, however, where time delay and group delay are equivalent: a pure time delay and an all-pass response. In both cases, magnitude response is flat. For the pure time delay:

$$G_d = -\frac{d\phi}{d\omega} = T_d \text{ for all values of } \omega$$

For this case, group delay and time delay are equal at all frequencies. This is true because the slope of the phase curve for a pure time delay is constant in frequency.

One of the reasons the all-pass case is important is because many popular crossover networks sum to an all-pass response. That is, when the outputs of the low-pass and high-pass networks are added together, they sum to a flat magnitude response, but the phase shift is not zero. The phase response is that of an all-pass network. The networks that do this include all odd-order Butterworth filters and the even-order Linkwitz-Riley (L-R) filters.

Phase response for the summed outputs of a 2kHz fourth-order L-R filter is plotted in *Fig. 6.49a*. Group delay is plotted in *Fig. 6.49b*. At very low frequencies, the woofer response is delayed relative to the tweeter by about 225µs or 0.225ms. This is true even if the driver acoustic phase centers are spatially aligned. It is a consequence of the phase response of the composite output of the two drivers.

Due to the rapid change in group delay in the crossover region, the time relationship between frequencies in this range is not preserved. The aural consequence of this effect is hotly debated, but experimental evidence to date concludes that it can only be heard with highly specialized test signals having steeply rising envelopes.

6.8 DETERMINING A DRIVER'S ACOUSTIC PHASE CENTER

The commonly used phrase "time alignment" of drivers is somewhat misleading. Heyser[11] has shown that there is no single number characterizing the spatial or temporal location of a driver. In regions where frequency response is *changing*, the signal-source locations for even a single frequency are distributed over a small region in space or, equivalently, over time. The frequency components of a sharply rising signal do not arrive at a listening location at one time. They are distributed over a narrow range of times.

To make this concept of a distributed source location clearer, consider the following simple example. *Figure 6.50* shows the frequency response of a first-order high-pass filter with a 100Hz cutoff frequency. Response is rising at 6dB/octave above 10Hz. The response slope begins to decrease near 100Hz and is essentially flat at 1kHz.

This same high-pass response can be obtained as the summed response of a straight wire and a first-order all-pass network. This equivalence is illustrated in *Fig. 6.51*. *Figure 6.52a* shows the phase response of the first-order all-pass. At very low frequencies the phase shift is +180°.

Thus the straight wire and all-pass outputs are 180° out of phase and the summed response is greatly attenuated. As frequency increases, the all-pass phase angle decreases, the summed signals are more in phase, and output rises. At very high frequencies, the all-pass and straight wire are in phase and response levels out as in *Fig. 6.50*.

You now see that the output of a high-pass filter can be viewed as the sum of two versions of the input signal, one arriving directly along the straight wire and a second delayed version of the input. *Figure 6.52b* is a plot of group delay for the first-order all-pass. Remember, group delay and time delay are equivalent for the all-pass. Notice that group delay is a straight line when plotted on a loglog scale.

At 10Hz the group delay is 3.18ms. The high-pass filter output at 10Hz consists of two signals of the same amplitude. The first signal arrives instantaneously at the output via the straight wire, while the second signal arrives 3.18ms later. At 100Hz the two signals arrive 31.8µs apart. At 1000Hz the two signals are 0.318µs apart. Thus, even for this very simple circuit, you cannot assign a single point in time to the arrival of a signal at the filter output.

Although driver signal arrivals are spread out in time, you can define an *acoustic phase center* which in practice is constant or nearly so over a usefully wide frequency range. The acoustic phase center of a driver is that location along the driver's axis which makes its phase response minimum phase.[12] This statement is most easily explained with an example.

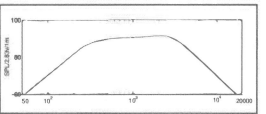

FIGURE 6.53a: Example 6.7: Magnitude response of midrange driver.

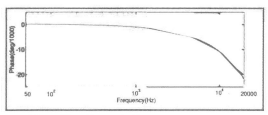

FIGURE 6.53b: Example 6.7: Unwrapped measured midrange driver phase.

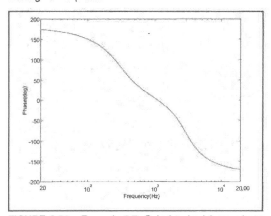

FIGURE 6.54a: Example 6.7: Calculated minimum phase response of midrange driver.

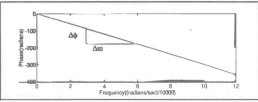

FIGURE 6.54b: Example 6.7: Excess phase response in midrange measurement.

6.8.1 EXAMPLE 6.7: DETERMINING DRIVER ACOUSTIC PHASE CENTER

This is a hypothetical example using an idealized driver to illustrate the procedure. It measures the impulse response of a midrange driver using the MLS technique and computes the frequency and phase response resulting from that measurement. The DUT is assumed to be mounted flush on a large baffle. The microphone is placed on the driver axis exactly 1m out from the baffle surface.

The measured frequency response is shown in *Fig. 6.53a* The driver −3dB points are at about 300 and 3000Hz. Response rises gently within the passband and falls off at 12dB/octave above 3000Hz and below 300Hz. The unwrapped measured phase response is shown in *Fig. 6.53b*. The phase angle equals 22,000° at 20kHz!

From the earlier low-pass filter example, you would expect a phase angle of no more than

113

−180° at this frequency. The very large measured phase angle is the excess phase buildup caused by the time delay between the application of the impulse to the DUT and its arrival at the microphone 1m away.

The minimum phase response associated with the measured frequency response in *Fig. 6.53a* can be computed with the Hilbert transform. This minimum phase response is shown in *Fig. 6.54a*. It starts out at +180° at low frequencies and winds up at −180° at high frequencies, just as you would expect from the measured magnitude response slopes.

Subtracting the computed minimum phase from the total measured phase results in the excess phase plotted in *Fig. 6.54b*. The excess phase is plotted in radians on a linear radian frequency scale. The resulting straight line shows that the excess phase is indeed due to a pure time delay. The slope of this line is −2.965ms. Use 344m/s for the speed of sound to compute a distance, d:

$$d = 2.965 \times 10^{-3} \times 344 = 1.02m = 102cm$$

The acoustic phase center of this hypothetical midrange is 102cm away from the microphone at a location which is also 2cm behind the baffle surface.

In practice, the absolute acoustic phase center is not of interest. For design purposes in multiple driver systems, you want to know the phase center offsets of the drivers. One driver, usually the tweeter, is taken as a reference. Then the phase center offsets of the mid and/or woofer relative to the tweeter are determined. With this approach, any uncompensated delays in the measurement system cancel out. You will see real examples of this procedure in Chapter 7.

REFERENCES

1. J.M. Berman and L.R. Fincham, "The Application of Digital Techniques to the Measurement of Loudspeakers," *JAES*, Vol. 25, pp. 370–382, June 1977.

2. L.R. Fincham, "Refinements in the Impulse Testing of Loudspeakers," *JAES*, Vol. 33, No. 3, pp. 133–140, March 1985.

3. R.A. Roberts and C.T. Mullis, *Digital Signal Processing*, Addison-Wesley, Reading, MA, 1987.

4. J.G. Proakis and D.G. Manolakis, *Introduction to Digital Signal Processing*, Macmillan Publishing Company, New York, 1988.

5. J.W. Cooley and J.W. Tukey, "An Algorithm for the Machine Computaion of Complex Fourier Series," Math. Comp., Vol. 19, pp. 297-301, April 1965.

6. Private communication with Richard Campbell, Bang-Campbell Associates, Woods Hole, MA, 1997.

7. F.J. Harris, "On the Use of Windows for Harmonic Analysis with the Discrete Fourier Transform," Proc. IEEE, Vol. 66, pp. 51–83, Jan. 1978.

8. R.J. Kos, "The Use of a Maximum-Length Sequence as an Excitation Signal to Acquire a System Impulse Response," Senior Project, Worcester Polytechnic Institute, April 25, 1991.

9. D. Rife and J. Vanderkooy, "Transfer-Function Measurements with Maximum-Length Sequences," *JAES*, Vol. 37, No. 6, pp. 419–443, June 1989.

10. B. Waslo, "The IMP Goes MLS," *Speaker Builder*, 6/93.

11. R.C. Heyser, "Determination of Loudspeaker Arrival Time, Parts I, II & III, *JAES*, Vol. 19, Nos. 9, 10, & 11, October, November, & December, 1971.

12. S.P. Lipshitz, T.C. Scott, and J. Vanderkooy, "Increasing the Audio Measurement Capability of FFT Analyzers by Microcomputer Postprocessing," *JAES*, Vol. 33, No. 9, September 1985, pp. 626–648.

CHAPTER 7

LOUDSPEAKER TESTING WITH PC-BASED ACOUSTIC DATA ACQUISITION SYSTEMS

7.1 INTRODUCTION

There are many PC-based acoustic data acquisition and analysis systems on the market today. These systems comprise two basic elements—hardware in the form of a plug-in card and software. Highly sophisticated systems, such as DRA Laboratory's MLSSA and SYSid Labs' SYSid, costing $3500–4000, have dedicated cards optimized for acoustic data acquisition. At the other end of the price scale are acoustic data-acquisition and analysis software packages which use your existing sound card for the hardware.

All of the examples you will see in this chapter have been generated with the MLSSA or CLIO systems. CLIO is made by Audiomatica, SRL, Italy. It falls between the two extremes discussed above. It has a dedicated board, but is priced in the $1100–1200 range. Although I would not hesitate to recommend these systems, this chapter is not an advertisement for either. Nor is it a contest between them. They have differing capabilities and operational philosophies.

By showing you results from two systems, I hope you will gain a better appreciation for the capabilities and power of PC-based acoustical and electrical measurement systems in general. There will be few if any step-by-step procedures given in this chapter. The emphasis will be on describing the wealth of new information these systems provide and how to interpret and use it.

This chapter starts with a brief description of the capabilities of the MLSSA and CLIO systems in Section 7.2. Then examples of impedance and capacitance and inductance measurements are presented in Section 7.3. Measurement of driver T/S parameters is covered in Section 7.4.

In Section 7.5 I discuss measurement types provided by CLIO and MLSSA for loudspeaker design and evaluation describing both time-domain and frequency-domain measurements. The full set of measurements needed for loudspeaker design are then covered with several examples in Section 7.6, while measurements for performance evaluation are described in Section 7.7. The chapter ends with some additional interesting examples which do not fit the two previous categories, but do highlight other properties or useful capabilities of these systems such as room/speaker equalization.

7.2 THE MLSSA AND CLIO SYSTEMS

7.2.1 MLSSA

Hardware and software functionality of the MLSSA system are summarized in *Table 7.1*. Only those features directly relating to loudspeaker measurement and analysis are covered in the table. The MLSSA

hardware consists of an 8-bit full-slot board with one 12-bit A/D converter and two autoranging four-pole low-pass analog antialiasing filters. The filters are normally cascaded for an overall 8-pole response.

Autoranging occurs in 1dB steps to get maximum dynamic range from the 12-bit converter. The antialiasing filter-response characteristic can be set to Bessel, Butterworth, or Chebychev under software control. An 8-pole Chebychev filter combined with a 125kHz sample rate yields an upper analysis bandwidth of 40kHz.

MLSSA operates in two basic modes: the time domain and the frequency domain. In the time domain mode, pulse, step, and MLS signals are provided for system excitation. MLS lengths of 4095, 16383, 32768, and 65535 (N = 12, 14, 15, and 16) are provided. The impulse response is the basic measurement performed with MLSSA.

All other data is derived from this measurement through software. Derived data includes step response, the Energy-Time Curve (ETC), and a waterfall plot of Cumulative Spectral Decay (CSD). Don't be too concerned if some of the preceding terms are unfamiliar to you. Many examples later in the chapter will make them clear.

MLSSA will also operate as a digital oscilloscope in the time-domain mode. Signals captured in the time domain can then be processed with many of the frequency-domain functions described below.

The transition from the time domain to the frequency domain is accomplished via an FFT, whose lengths range from 16 samples to 64K samples. There are multiple FFT modes. The three of most interest to us include the impedance mode, the transfer-function mode, and the sensitivity mode. The impedance mode computes impedance magnitude and phase of the device. This can then be displayed in magnitude and phase plots, Bode plots, or Nyquist plots.

TABLE 7.1
MLSSA HIGHLIGHTS

Hardware:	8-bit full-slot card
	1 12-bit R-2R ladder A/D
	2 four-pole autoranging analog antialiasing filters (Bessel, Butterworth, Chebychev, G = −14 to +54dB)
	sample rates: 1kHz–125kHz
	analysis bandwidths to 40kHz
Signal types:	pulse, step, and MLS, N = 12, 14, 15, 16
Time domain data:	Impulse, step, ETC, waterfall (CSD), scope mode
Frequency domain data:	FFTs of length (L = 16 to 64K)
	Impedance mag. and phase, T/S parameters with SP option
	Frequency response: Mag & Phase, minimum phase, group delay excess phase, excess group delay, driver acoustic center
	Nyquist, Bode, Real & Imaginary display modes L/C values, THD distortion with external sine source

---

You will see examples of these plots later in the chapter. MLSSA will also calculate inductance and capacitance in the impedance mode.

The transfer-function mode computes an input/output transfer-function magnitude and phase. The sensitivity mode is similar, but it is specialized for loudspeaker response. In this mode microphone calibration data is used to compute response in dB SPL relative to a power or voltage level at a fixed distance. You can set power or voltage and distance parameters for the measurement in the time domain screen. MLSSA will also measure harmonic distortion with an external sine-wave oscillator supplying the input signal.

7.2.2 THE CLIO SYSTEM

Hardware and software functionality of the CLIO system is summarized in *Table 7.2*. The CLIO hardware consists of an 8-bit half-slot board with two 16-bit sigma-delta D/A and A/D channels. The two D/A channels generate various test signals, while the two A/D channels acquire the time domain response to those signals. CLIO provides sample rates of 3.2, 12.8, and 51.2kHz. The highest sample rate results in an analysis bandwidth of 20kHz.

Signals generated by CLIO include sine waves; frequency-stepped sine waves; tone bursts; two-tone, multi-tone, and all-tone signals; white noise; pink noise; and an MLS signal. CLIO has seven operating modes: spectrum analysis (the FFT mode), MLS, Sinusoidal, RTA, Acoustics, Tools and QC. The first four modes are of primary interest. The FFT mode is useful for general spectrum analysis and distortion measurement. The MLS mode gets us the time-domain impulse response, frequency response, the ETC, and CSD.

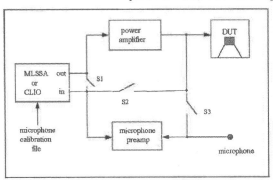

FIGURE 7.1: Possible calibration scenarios for MLSSA and CLIO.

The Sinusoidal mode produces frequency-domain response and impedance data using stepped sine-wave sweeps. Step spacings of 1/3, 1/6, 1/12, 1/24, and 1/48 octaves are provided. Upper and lower limits on the frequency range of the sweep are under operator control. Second- and third-harmonic distortion performance and T/S parameters are also measured in this mode. The RTA mode generates 1/3 octave response plots with pink noise. This mode is also useful for loudspeaker/room equalization. The Tools mode adds a two-channel scope and L/C measurement capability.

Both MLSSA and CLIO have additional acoustical and electrical measurement capabilities which are not usually needed for loudspeaker analysis and evaluation. MLSSA has a full set of functions for evaluating acoustical spaces such as large auditoria and concert halls. It also has several built-in quality-control functions for production monitoring. CLIO has a full menu of QC functions with its optional QC package. I will not discuss these capabilities here.

7.2.3 MLSSA AND CLIO CALIBRATION

Both MLSSA and CLIO must be calibrated to attain their full accuracy potential. In fact, CLIO will not perform any measurements until a full set of self-calibration files are present. Three possible calibration scenarios are shown in *Fig. 7.1*. All involve a loop back from output to input of all or part of the signal-processing chain.

Closing switch S1 with S2 and S3 open connects the output signal back to the input to perform a self-calibration, which measures the response of the internal signal generation and data-acquisition path. It accounts for any variation in signal level with frequency and for the frequency and phase response of any internal amplification and the antialiasing filter. This procedure is fully automatic in CLIO and produces six calibration files—three MLS calibration files corresponding to the three sample rates plus sine, RTA, and level calibration files. Depending on which mode CLIO is in, one of these files will always be used to correct the measured data for CLIO's internal response variations.

CLIO can also apply an additional correction file to any measurement through its Reference command. For example, closing S2 with S1 and S3 open will allow CLIO to measure the power amplifier frequency and phase response, which can be saved as a reference file and used to correct measured loudspeaker response for any frequency-response variations or phase shift in the power amp. Closing S3 with S1 and S2 open will include the microphone preamp in the calibration file. Finally, CLIO will also accept a microphone calibration file to correct acoustic measurements if one is available.

Because CLIO has only six distinct operating modes requiring calibration, automating the self-calibration process is relatively easy. The situation with MLSSA is quite different. With sample rates extending from 1kHz to 125kHz, three

TABLE 7.2
CLIO HIGHLIGHTS

Hardware:	8-bit half-slot card Analyzer: 2 channel 16-bit sigma-delta A/D Generator: 2 channel 16-bit sigma-delta D/A Sample rates: 3.2, 12.8, and 51.2kHz Analysis bandwidth to 20kHz
Signal types:	sine, swept sine, tone burst, two-tone, multitone, all tone, white noise, pink noise, MLS
Modes:	FFT spectrum analysis, MLS, Sinusoidal, RTA, Acoustics, Tools
Time domain data:	Impulse response, ETC, CSD waterfall, scope
Frequency domain data:	Impedance mag & phase, frequency response mag & phase, minimum phase, driver delay, THD, IMD, RTA, T/S parameters

different antialiasing filter types having selectable bandwidths ranging from 1kHz–40kHz, and FFT sizes running from 16 samples to 64K, the number of possible self-calibration files for MLSSA is extremely large.

To limit the number of calibration files, I have standardized on three sample rates–4kHz, 62kHz, and 125kHz. I use a Bessel antialiasing filter with the 4kHz rate and Chebychev filters with the 62 and 125kHz sample rates. These sample rates and filter types provide analysis bandwidths of 1, 20, and 40kHz, respectively.

The calibration files are made with 16K FFTs. MLSSA will convert the 16K files to any other FFT size through its Reference-Import command. When necessary, additional calibration files are easily generated for special situations such as those requiring inclusion of power amplifier and/or microphone preamp compensation. MLSSA will also apply microphone corrections through the Reference-Auxiliary command.

7.3 IMPEDANCE AND L/C MEASUREMENTS

Both MLSSA and CLIO perform impedance measurements, but they differ in the test stimulus used. MLSSA excites the impedance to be tested with a broadband MLS signal, while CLIO uses a stepped sine-wave sweep. Both use the voltage divider technique discussed in Chapter 2. It will be interesting to compare results from both systems, but first let's revisit the voltage divider technique for measuring impedance.

7.3.1 A NEW LOOK AT THE VOLTAGE-DIVIDER TECHNIQUE

I introduced the voltage-divider technique for measuring impedance in Section 2.6.2. The voltage-divider diagram, Fig. 2.16, is repeated in *Fig. 7.2* for easy reference. The voltage-divider equation, Equation 2.20, is repeated below:

$$V_{out} = \left(\frac{Z}{R+Z} \right) V_{gen} \qquad [7.1]$$

Recall from Chapter 2 that this equation cannot be solved exactly for the unknown impedance, Z, without phase information. An approximate solution for the *magnitude* of Z, denoted |Z|, was described, however, by making R much larger than |Z|.

Rewriting Equation 7.1 in a slightly different form gives us:

$$\left(\frac{V_{out}}{V_{gen}} \right) = \frac{Z}{R+Z} \qquad [7.2]$$

The ratio, V_{out}/V_{gen}, is simply the voltage-divider transfer function, which both MLSSA and CLIO measure. This ratio is a complex quantity. That is, it has both real and imaginary parts, or, equivalently, it has both magnitude and phase. Both systems measure the magnitude and phase of V_{out} relative to V_{gen}. With both magnitude and phase information, it is now possible to solve Equation 7.2 for Z exactly, without

the limitation on R. A little algebra gives us a solution in terms of the measured transfer function:

$$Z = \frac{\left(\dfrac{V_{out}}{V_{gen}} \right) R}{1 - \dfrac{V_{out}}{V_{gen}}} \qquad [7.3]$$

7.3.2 MEASURING IMPEDANCE WITH CLIO AND MLSSA

The setups used by MLSSA and CLIO for measuring impedance are basically identical (*Fig. 7.3*). The resistor, R, is internal to both systems. CLIO is a dual-channel system with both A and B input and output channels. Impedance measurements using the internal resistor are made with the A channel. Two connections are made between the hardware and the unknown impedance. The output connection applies a voltage stimulus to the unknown impedance through the internal resistor. The input

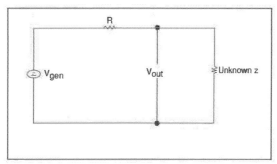

FIGURE 7.2: The voltage divider for impedance measurement.

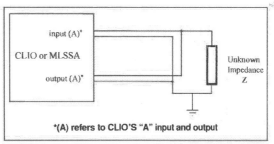

FIGURE 7.3: CLIO and MLSSA impedance measurements with internal resistor.

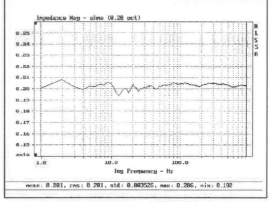

FIGURE 7.4: MLSSA impedance calibration with 8.20Ω resistor.

connection is used to measure the voltage magnitude across the unknown impedance and its phase angle relative to the original stimulus. This data provides the transfer function, which is then used to compute Z using Equation 7.3.

For greatest accuracy MLSSA should be calibrated in the impedance mode. One of the input variables in this mode is the value of R. The internal resistor in MLSSA is a 75Ω/5% resistor. Before producing the impedance data shown throughout this chapter, I calibrated MLSSA using an 8.2Ω/1% noninductive power resistor.

Figure 7.4 shows the results of the calibration process. MLSSA will calculate a number of statistics on the measured spectrum. The statistic of interest here is its mean (average) value over the analysis bandwidth, which in this instance is 1Hz to 1kHz.

R was varied in small steps about its nominal 75Ω value in a sequence of trials until an average value of 8.20Ω was obtained. The value of R giving the correct answer was 75.2Ω. The plot of resistance versus frequency looks pretty ragged until you realize that the vertical scale divisions are only 0.01Ω. The statistics printed below the plot show an average value of 8.201Ω, with peak deviations from the mean of +0.005Ω and −0.009Ω, which is less than 1 part in 820.

7.3.3 MEASURING INDUCTANCE AND CAPACITANCE

Both CLIO and MLSSA measure inductance and

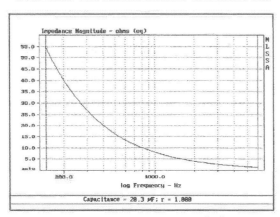

FIGURE 7.5: Measured impedance of 20µF capacitor (150Hz–6kHz).

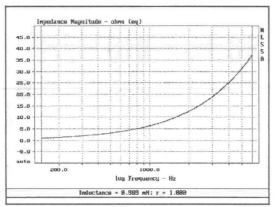

FIGURE 7.6: MLSSA measured impedance of 1.0mH coil (150Hz–6kHz).

**TABLE 7.3
CLIO L/C MEASUREMENT RESULTS**

Measured value	Measurement Frequency
1.00mH	1995.26Hz
19.86µF	794.33Hz

**TABLE 7.4
MLSSA L/C MEASUREMENT RESULTS**

Component Type	Frequency Range	Value(r)
C	1–4kHz	20.1µF(1.000)
C	150Hz–6kHz	20.3µF(1.000)
C	20Hz–20kHz	26.7µF(0.980)
L	1–4kHz	0.994mH(1.000)
L	150Hz–6kHz	0.989mH(1.000)
L	20Hz–20kHz	0.915mH(0.980)

capacitance by first measuring the impedance of the component under test. Beyond that, their approaches are quite different. CLIO uses a *single* frequency sine wave to measure L and C, automatically selecting the test frequency for greatest accuracy. MLSSA uses an MLS signal to measure impedance over a user-selected frequency *range* and then fits a component model to the measured data.

Let's look at typical measurements of inductance and capacitance with both systems. I pulled a high-quality 20µF polypropylene capacitor and a 1mH, 16-gauge perfect lay coil out of my parts bin for testing. Both parts are 5% tolerance components.

Component values are measured with CLIO by first entering the "Tools" menu and selecting the "LCmeter" option. *Table 7.3* lists the measured values and the frequency selected by CLIO for the measurement. The measured values are at or very close to their labeled values and much better than a 5% tolerance. Inductance and capacitance measurements with CLIO take less than 10s.

The procedure used by MLSSA to measure capacitance and inductance is different and perhaps unique. First, the component impedance is measured over a wide frequency range with an MLS stimulus. Next, the frequency range over which an *effective* capacitance or inductance value is desired is selected by setting a marker and cursor on the impedance plot. The MLSSA program least squares-fits a reactance model to just the reactive (i.e., imaginary) portion of the impedance curve over the selected frequency range.

This approach ensures that the calculated capacitance or inductance is not contaminated by any series resistance in the test leads. The models fitted to the data are the well-known equations for capacitive and inductive reactance given as:

$$X_C = \frac{1}{2\pi fC} \quad \text{and} \quad X_L = 2\pi fL$$

where X_C and X_L are capacitive and inductive reactance, respectively, and f is frequency. MLSSA adjusts the value of C or L to best fit the measured reactance over the selected frequency range in a least-squares sense. This is a very different concept from the single frequency value given by CLIO or by RLC bridges in general. The entire process of measuring the impedance and selecting the frequency range typically takes about 30s.

The capacitor and inductor values were next measured with MLSSA. Because MLSSA calculates a best-fit model over a frequency range while CLIO computes L and C at a single frequency, you should not expect the two systems to get the same values. The impedance magnitude of the 20μF capacitor measured with MLSSA is shown in *Fig. 7.5*.

I made three capacitance calculations with this data. For the first calculation I selected a frequency range of 1–4kHz, which brackets the frequency selected by CLIO from above and below by one octave. For this frequency range MLSSA calculated a value of 20.1μF.

To select the second frequency range, I assumed the capacitor (and inductor) would be used in a 300Hz–3kHz midrange crossover network. The active frequency range for this network will span a region from one octave below crossover to one octave above; that is, from 150Hz to 6kHz. Over this range MLSSA calculates an effective capacitance value of 20.3μF. The third frequency range included the entire audio band from 20Hz–20kHz and produced a capacitor value of 26.7μF.

The impedance magnitude of the 1mH coil measured with MLSSA is shown in *Fig. 7.6*. Values of capacitance and inductance calculated by MLSSA for the three frequency ranges are listed in *Table 7.4*. The term (r), heading the third column, is a correlation coefficient which is a measure of how

well the model correlates with the measured data. It has a maximum value of one. Values less than one indicate a poorer fit.

Examining *Table 7.4*, notice that the model-fit capacitance increases as the frequency span increases, while the model-fit inductance decreases. Furthermore, r falls below one for the largest frequency span which comprises the entire audio band. This is happening because the capacitor is not an ideal capacitor. Nor is the inductor an ideal inductor. Real-world capacitors have small amounts of series inductance and resistance and parallel leakage resistance. Real-world inductors have series resistance and inter-winding shunt capacitance.

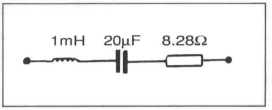

FIGURE 7.9: Series RLC circuit.

You can see the effect of these parasitic elements in the impedance data. Impedance magnitude and phase of the 20μF capacitor measured with CLIO are plotted in *Fig. 7.7*. The left-hand side vertical scale, which is impedance magnitude, is logarithmic on this plot. On a logarithmic scale the impedance magnitude of an ideal capacitor will plot as a straight line with a negative slope and a constant phase of –90°.

Note, however, that the measured phase angle begins to rise (become less negative) above

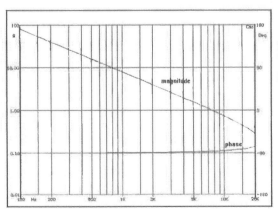

FIGURE 7.7: Measured impedance of 20μF capacitor.

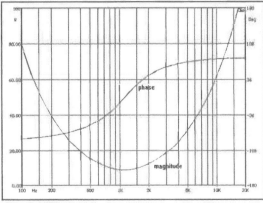

FIGURE 7.10: CLIO impedance measurement.

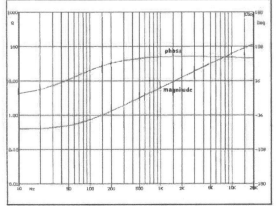

FIGURE 7.8: 1mH coil impedance magnitude and phase.

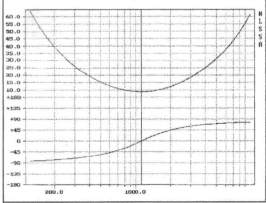

FIGURE 7.11: MLSSA measured impedance of RLC circuit.

2kHz and the magnitude plot begins to show significant downward curvature above 8kHz. The measured impedance magnitude and phase at 20kHz are 0.3Ω and $-81.2°$, respectively. For comparison, values for an ideal $20\mu F$ capacitor are 0.4Ω and $-90°$. The decreased reactance at the upper end of the audio band causes MLSSA to calculate a larger value of capacitance for the model fit. The decreased reactive component is probably caused by small values of series inductance and resistance. The series inductance is a byproduct of the winding of the foil-dielectric sandwich making up the capacitor.

Figure 7.8 shows the impedance magnitude and phase of the 1mH inductor. The impedance magnitude should be a straight line with a positive slope, and phase should be constant at $+90°$. The coil impedance magnitude and phase

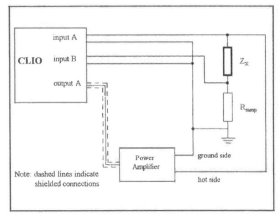

FIGURE 7.12a: Constant voltage mode impedance measurement using CLIO with a booster amplifier.

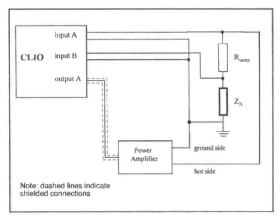

FIGURE 7.12b: Constant current mode impedance measurement using CLIO with a booster amplifier.

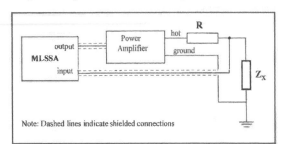

FIGURE 7.13: Measuring impedance with MLSSA and a booster amplifier.

at 10Hz are 0.38Ω and $10.3°$, respectively. This is very close to a pure resistance. The coil has a DC resistance (DCR) of 0.37Ω. An ideal 1mH coil has a reactive impedance magnitude of only $+0.063\Omega$ at 10Hz.

The coil DCR dominates the inductor's impedance at low frequencies. The coil only begins to look like an ideal inductor above 150Hz. Including more of the low-frequency impedance data in MLSSA's modeling calculation caused it to fit lower values of inductance to the data.

Here is one more interesting point on the coil. The coil's impedance phase never reaches $+90°$. It has a maximum value of $+87.2°$ at 2400Hz. Above that frequency it falls slowly to a value of $+83.2°$ at 20kHz, probably due to the shunting effect of interwinding capacitance. Although our primary interest will be in measuring the impedance of drivers and loudspeaker systems, the complex impedance data provided by CLIO and MLSSA shows a great deal about a component's electrical behavior.

I will end this section with one more example. The $20\mu F$ capacitor and 1mH coil were combined with an 8.28Ω resistor to form the series RLC circuit (*Fig. 7.9*). Using these label values for L and C, this circuit will have a series resonance at a frequency given by the formula:

$$f_{ser} = \frac{1}{2\pi\sqrt{LC}} = \frac{0.1592}{\sqrt{(1\times10^{-3})(20\times10^{-6})}} = 1125.7\text{Hz}$$

At this frequency inductive and capacitive reactances are equal and cancel, leaving a purely resistive impedance equal to the sum of the coil resistance and the 8.28Ω resistor, which is 8.65Ω.

The impedance of this RLC circuit was measured with both CLIO and MLSSA. The CLIO measurement was made with a sine wave sweep in 1/48 octave steps. The point of zero phase with the CLIO measurement shown in *Fig. 7.10* fell precisely on one of the frequency sample points at 1128.64Hz. This is about 0.25% higher than the value calculated above using label values.

The RLC circuit impedance magnitude and phase measured with MLSSA are shown in *Bode* plot format (*Fig. 7.11*). The point of zero phase did not fall in an FFT frequency bin.

The sample rate and sample size used in the MLSSA measurement resulted in an FFT bin width of 3.875Hz. Phase angles of $-0.5°$ and $+0.6°$ were measured at 1127.697Hz and 1131.572Hz, respectively. Notice that the two MLSSA data points bracket the CLIO measured value. For all practical purposes the two measurements agree. The impedance minima measured with CLIO and MLSSA were 8.68Ω and 8.65Ω, respectively. All in all, CLIO and MLSSA agreed to within 0.1%.

7.3.4 MEASURING IMPEDANCE AT HIGHER DRIVE LEVELS

When measuring impedance with its 100Ω internal resistor, CLIO automatically sets the sine-wave drive level to 0dBm, which is 0.775V RMS. The MLSSA manual clearly states that to prevent satu-

LOUDSPEAKER
TESTING WITH
PC-BASED
ACOUSTIC DATA
ACQUISITION
SYSTEMS

rating the on-board op amp the MLS signal level should be set no higher than 1V peak when using its 75Ω internal resistor to measure impedance.

Chapter 2 describes drivers as weakly nonlinear devices. Both impedance and T/S parameters change with increasing drive level. If you wish to measure impedance or T/S parameters at higher drive levels with either CLIO or MLSSA, an external amplifier is required.

In addition to the internal resistor mode, CLIO offers two other impedance measurement modes using an external amplifier: the constant-voltage and constant-current modes described in Chapter 2. These two modes are greatly facilitated by CLIO's two-channel capability. Setups for these modes are illustrated in *Figs. 7.12a* and *b*.

In the constant-voltage mode, channel A measures the voltage across the series combination of the unknown impedance, Z_X and the sampling resistor, R_{samp}. Channel B measures the voltage across R_{samp}. Both voltages are ground referenced. CLIO then finds the voltage across Z_X by differencing the A and B voltages. The current through Z_X is determined by dividing the B voltage by the value of R_{samp}. (The value of R_{samp} is input to CLIO prior to the measurement.)

Finally, the complex impedance is computed by dividing the voltage by the current. R_{samp} is chosen to be much smaller than Z_X to keep the voltage across it relatively constant. CLIO's constant voltage mode was used to generate the set of T/S parameters versus drive level given in Table 2.1.

Referring to *Fig. 7.12b*, the roles of R_{samp} and Z_X are reversed in the constant-current mode. R_{samp} is now selected to be much larger than Z_X to keep current relatively constant.

In the setup for measuring impedance at higher drive levels with MLSSA (*Fig. 7.13*), only the voltage-divider method is supported. A two-step calibration is required before impedance measurements can be made. The first step involves a loop back through the power amplifier to compensate the impedance measurement for the amplifier's frequency and phase response. This is done by closing switch S2 in *Fig. 7.1* with switches S1 and S3 open. The second step is similar to the impedance calibration with the internal resistor. The value of R input to MLSSA is changed in small increments until the measured resistance agrees with the value of a precision test resistor put in place of Z_X.

7.4 THIELE/SMALL PARAMETER MEASUREMENTS

7.4.1 THE "THREE-POINT" METHOD VERSUS MODEL FITTING

In Chapter 2 I described the "three-point" method for determining T/S parameters with driver impedance data. In this method, the impedance magnitude at three frequencies is measured and used to compute the T/S parameters. Recall that great care is required with this method to assure that the frequencies f_1 and f_2 are accurately measured, since the equations used to calculate the parameters are very sensitive to small errors in these frequencies. As discussed in Chapter 2, the three-point method

is also very susceptible to noise.

The three-point method assumes that the measured points lie on the curve of the simple theoretical model of driver impedance shown in Fig. 2.14. In this model the impedance curve has geometric symmetry about f_{SA}. This model does not include the effect of voice-coil inductance and frequency-dependent changes in voice-coil impedance which can distort the impedance curve. The impedance curve of a real driver rarely matches the theoretical model exactly.

CLIO improves upon the three-point method by

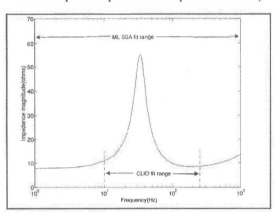

FIGURE 7.14: Frequency range of MLSSA and CLIO model fits.

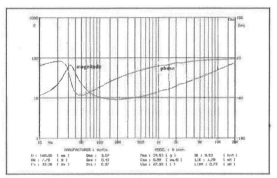

FIGURE 7.15: CLIO T/S parameters for 8″ driver.

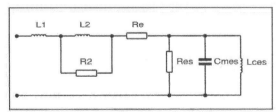

FIGURE 7.16: Driver electrical model used by MLSSA.

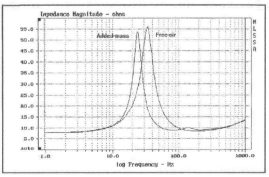

FIGURE 7.17: 8″ driver free-air and added-mass impedance curves.

least-squares-fitting a second-order model to the driver's measured impedance magnitude over a large number of frequency points spanning the region around resonance. MLSSA least-squares-fits a more sophisticated model to both the driver's impedance magnitude and phase, which not only covers the region around f_{SA}, but also carries the fit well beyond the first minimum above f_{SA}. The MLSSA model accounts for frequency dependent inductance and resistance.

Typical frequency ranges over which a model is fitted by both systems are illustrated in *Fig. 7.14*. In each case, a least-squares-fitting process finds a set of parameters for the chosen model that best fit the impedance data over a large number of points. The effect of a few noise-induced outliers is minimized by using many points. Once a model is fitted to the data, driver parameters are computed from the model.

7.4.2 MEASURING T/S PARAMETERS WITH CLIO

T/S parameters are measured with CLIO by first entering the Sinusoidal menu and then selecting the Parameters option. This selection produces the Impedance Response and Parameters screen (*Fig. 7.15*). T/S parameter estimates involve a three-step process. The first step is initiated by clicking on the Q radio button or striking the "Q" key. CLIO asks for the driver manufacturer, model number, and voice-coil resistance, R_E, which must be measured by independent means. CLIO then proceeds to measure driver impedance over the full audio range using a stepped sine-wave stimulus.

Once impedance data is available a second-order model is fit to the impedance curve in a frequency

range about f_{SA} (*Fig. 7.14*). From this model and higher frequency impedance data, CLIO computes and displays driver resonant frequency and Qs and voice-coil inductance at 1kHz and 10kHz. *Note that all of the precautions regarding driver mounting and ambient noise outlined in Chapter 2 must be observed.*

You can stop at this point or continue on to measure M_{MS}, C_{MS}, V_{AS}, and Bl by selecting either the Added Mass or Known Box option. CLIO asks for the added mass in grams or the box volume in liters and the driver diameter. Once this information is provided, CLIO automatically performs a second impedance measurement with the added mass or with the driver placed in the known volume and calculates the additional parameters.

Fig. 7.15 shows the measured impedance magnitude and phase and the results of a full set of T/S parameters calculated by CLIO for an 8″ stamped-frame woofer using the added mass technique. These results will be compared against results obtained with MLSSA on the same driver in the next section. The entire measurement process takes about two minutes.

7.4.3 MEASURING T/S PARAMETERS WITH MLSSA

Recall from Chapter 2 that the "three-point" method for measuring T/S parameters is based upon a simplified model of voice impedance which ignores voice-coil inductance. This impedance has geometric symmetry about f_{SA}. Voice-coil inductance skews the impedance curve about f_{SA}, destroying that symmetry. Chapter 2 also explained that both voice-coil resistance and inductance are frequency dependent. This led to the frequency-dependent voice-coil impedance model shown in Fig. 2.11.

MLSSA uses a seven-parameter, frequency *independent* model to approximate this frequency-dependent behavior. MLSSA's model is shown in *Fig. 7.16*. Compare this with the more general model shown in Fig. 2.11. The L2/R2 parallel combination in *Fig. 7.16* approximates the frequency-dependent behavior of the voice-coil inductance and resistance. It tends to be an excellent model

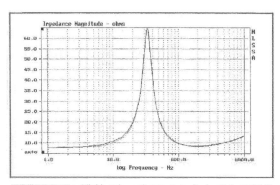

FIGURE 7.18a: 8″ driver impedance magnitude and model fit (dotted).

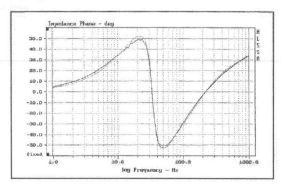

FIGURE 7.18b: 8″ driver impedance phase and model fit (dotted).

**TABLE 7.5
COMPARISON OF T/S PARAMETERS
DETERMINED WITH MLSSA AND CLIO**

Parameter	MLSSA	CLIO	MLSSA(2)
$R_E(\Omega)$	7.66*	7.70**	7.66
f_{SA} (Hz)	33.22	32.29	32.28
Q_{MS}	3.40	3.07	2.94
Q_{ES}	0.45	0.42	0.43
Q_{TS}	0.40	0.37	0.37
RMSE-free	0.466	****	0.374
M_{MS}(gm)	24.00	24.53	22.93
C_{MS}(mm/N)	0.96	0.99	0.106
V_{AS} (ltr)	66.19	67.95	73.39
Bl(N/A)	9.19	9.53	8.76
SPL(dB)	89.8	****	89.8
RMSE-load	0.987	****	0.626

* measured by MLSSA   ** measured with a
DCR bridge   **** Not supplied by CLIO

LOUDSPEAKER
TESTING WITH
PC-BASED
ACOUSTIC DATA
ACQUISITION
SYSTEMS

for frequencies up to about two octaves above the first impedance minimum above f_{SA}.

The MLSSA system least-squares-fits this model to both the magnitude and phase of the measured driver impedance. T/S parameters can be calculated from the model parameters. MLSSA will measure the DC resistance of the voice coil to within 0.01Ω if the entire measuring path is DC coupled. If the measurement path is AC-coupled, MLSSA will accept an independent measurement of voice-coil DCR. Finally, MLSSA can extrapolate the fitted model to DC and report out an "estimated" voice coil DCR, but this is the least accurate of the three options.

As with all other previously discussed methods, a single impedance measurement does not provide enough information to compute all of the T/S parameters. In addition to the added-mass and closed-box measurements already discussed, MLSSA offers a third option called "Fixed M_{MD}." If the value of the moving mass is known or can be measured by other means, it can be input to the program. In this case only one impedance measurement is needed to compute all T/S parameters of interest. The two-pass measurement with MLSSA takes about one minute to complete.

Parameters for the 8″ stamped-frame driver first measured with CLIO were measured again with MLSSA, also using the added-mass technique. The measured free-air and added-mass impedance curves used to estimate the model parameters are plotted in *Fig. 7.17*. MLSSA has a very interesting visual check of this process.

The estimated model parameters are used to synthesize a driver impedance curve, which can then be plotted against the measured curve for comparison. Measured impedance magnitude

and phase are compared against the synthesized impedance magnitude and phase in *Figs. 7.18a* and *b*. The fit of the synthesized model to measured data is very good, especially in the region around resonance.

The MLSSA-derived T/S parameters for the 8″ driver are listed in *Table 7.5*, along with those values previously determined with CLIO. The "RMSE-free" and "RMSE-loaded" numbers in the MLSSA column measure how well the MLSSA model fits the measured free-air and mass-loaded impedance data. "RMSE" stands for root-mean-square error and in this application it has the units of ohms. A value less than 10% of the mean speaker impedance indicates a good fit. For example, a value of 0.8 or less would be considered a good fit for an 8Ω speaker. CLIO does not provide a goodness-of-fit measure.

Referring to *Table 7.5*, notice that the resonant frequency and Qs measured with MLSSA are all higher than those obtained with CLIO. The difference is caused by the different test signals used and the effective drive level they produce. Recall the

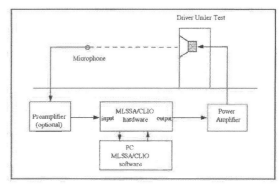

FIGURE 7.20: MLSSA/CLIO setup for response testing.

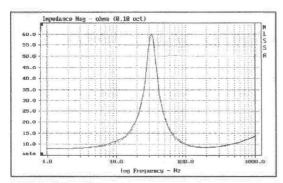

FIGURE 7.19a: 8″ driver impedance magnitude and model fit (dotted).

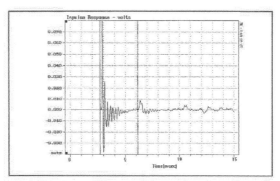

FIGURE 7.21: Impulse response of 7″ driver.

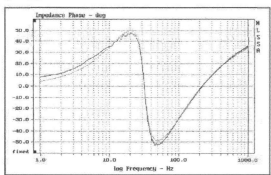

FIGURE 7.19b: 8″ driver impedance phase and model fit (dotted).

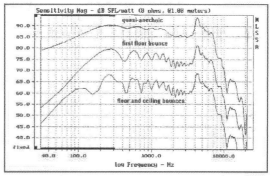

FIGURE 7.22: Frequency response versus time window width.

results of Table 2.1, which showed that f_{SA} and all Qs fell in value with increasing drive levels or equivalently with increasing peak cone excursion.

CLIO performs the T/S tests with a swept sine wave having an rms value of 0.775V, which is a peak value of 1.10V applied to the DUT through a 100Ω internal resistor. This peak value is applied at every frequency used in the test. MLSSA tests with the MLS signal, which is essentially a broadband noise. For the 4kHz sample rate chosen, the peak voltage applied to the internal resistor in any 1Hz bandwidth is only about 0.024V.

The resonant filter formed by the voice-coil impedance changes the character of the driving voltage across the voice coil from the binary-valued MLS signal to a Gaussian signal with an effective peak value of roughly 0.33V. (This is true for the 8″ driver under discussion. For a more general development of the equivalence of MLS and sine wave testing, see Reference 2.)

In order to obtain T/S parameters comparable to those produced by CLIO, the MLS drive level must be increased by a factor of 1.1/0.33, which is

3.33V. To obtain this level with MLSSA an external amplifier must be used with the setup shown in *Fig. 7.13*. T/S parameters obtained with MLSSA at this drive level are listed in *Table 7.5* under the column heading MLSSA(2). Note that f_{SA} has dropped relative to the value obtained with MLSSA at the lower drive level, and f_{SA} and the Qs are now comparable to the CLIO values.

However, the agreement of V_{AS} and M_{MS} with CLIO values is now somewhat worse. It is quite possible that exact equivalence of the two test methods cannot be attained. The impedance magnitude and phase computed using the estimated T/S parameters are compared against the measured curves in *Figs. 7.19a* and *b*. This model fit is excellent and even better than the fit obtained at the lower drive level.

Whether you use MLSSA or CLIO to measure T/S parameters, it is important to remember that they are "small signal" parameters. Drive levels should always be kept to the lowest level that provides reliable data.

7.5 MEASUREMENTS FOR LOUDSPEAKER SYSTEM DESIGN AND EVALUATION

In this section I will describe in general terms the more commonly used measurements provided by MLSSA and CLIO for loudspeaker design and evaluation. The measurement types fall into two general categories: time domain and frequency domain. Detailed examples of each measurement type will be given in Sections 7.6 and 7.7.

7.5.1 LOUDSPEAKER/DRIVER QUASI-ANECHOIC RESPONSE

Both MLSSA and CLIO use an MLS stimulus to

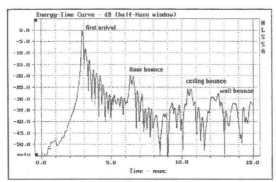

FIGURE 7.23: Energy-time plot for 7″ driver.

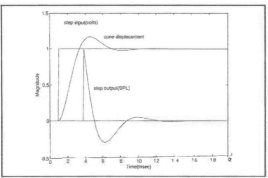

FIGURE 7.24: Step response of an ideal loudspeaker.

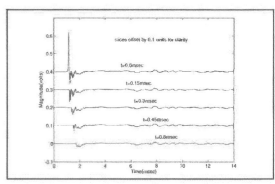

FIGURE 7.26: Five slices of impulse response used to compute the CSD.

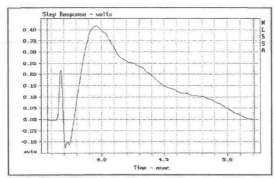

FIGURE 7.25: Step response of two-way MTM system.

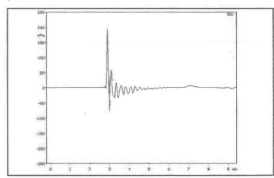

FIGURE 7.27: CLIO's impulse response of 7″ driver.

measure loudspeaker system or driver impulse response. In general this response includes the room reflections. The setup for measuring impulse response is shown in *Fig. 7.20*. It is similar to that shown in Fig. 4.1, with the sine-wave source and AC voltmeter replaced by a MLSSA or CLIO card under software control. All of the environmental issues affecting measurement accuracy discussed in Chapter 4 apply here, but as you will see shortly, you will be able to mitigate many of them.

The impulse response of a 7″ driver mounted in a 15 ltr vented enclosure was measured with MLSSA. The microphone was placed on the driver axis at a height of 38″ and 36″ away from the enclosure front baffle. This placement resulted in a floor bounce path length of 84″. The first 15ms of the measured impulse response is shown in *Fig. 7.21*.

From the figure you can see the first arrival of sound energy occurring at roughly 2.7ms. This is followed by a decaying oscillatory response, which falls to a very low level by 6ms. At 6.2ms a second, smaller burst of sound energy arrives. This is the floor bounce. A third arrival at about 10.5ms is a reflection from the ceiling of the test room. Finally, there is a fourth negative going arrival just beyond 12ms from a side wall.

Now you can begin to appreciate the power of time-domain data capture. Room reflections are clearly evident in the time-domain data. Placing a cursor and marker at 2.7 and 6.1ms, respectively, tells MLSSA (CLIO works in an identical manner) to analyze only the data that is free of all reflections. This interval includes only the direct arrival data.

The frequency response computed from this time interval is called the *quasi-anechoic* response. If the marker is now moved out to 6.6ms, the effect of the first floor bounce will be included in the calculated frequency response. Similarly, moving the marker out to 11ms will include the effect of both the floor and ceiling bounces in the frequency response.

The frequency responses corresponding to these three time intervals are shown in *Fig. 7.22*. Each plot is offset by 12.5dB for easier interpretation. The quasi-anechoic response covers only the first 3.4ms of the impulse response. Thus the lowest valid frequency in the FFT and also its true resolution is $1/.0034s$, or 294Hz. This is why the quasi-anechoic curve falls off rapidly below 300Hz. The data interval analyzed has no energy below that frequency.

The response curve including the floor bounce shows the classic comb-filtering effect first described in Chapter 4. This curve is valid down to about 250Hz. The lowest curve includes the effect of both ceiling and floor bounces with the lowest valid frequency and resolution of 90Hz. Below 1kHz the effect of the two bounces on the frequency response is interlaced, and they often interfere with each other.

7.5.2 THE ENERGY-TIME CURVE

The energy-time curve (ETC) gives another view of the temporal distribution of sound arrivals. The ETC is derived from the impulse response. Formally, the ETC is the envelope of the analytic signal. The analytic signal in turn is a complex quantity with a real part equal to the impulse response and an imaginary part equal to the Hilbert transform of the impulse response.

The analytic signal was first described by Gabor.[3] It is very useful in the analysis of communications systems. Heyser[4] introduced the analytic signal to the loudspeaker design community as a way to evaluate time of arrival.

Strictly speaking, the ETC does not represent sound energy.[5] Think of it somewhat loosely as a measure of the magnitude of the impulse response. Being always positive, the ETC removes any confusion that might be caused by negative going ripples in the impulse response. Both MLSSA and CLIO compute the ETC from the measured impulse response.

The ETC corresponding to the impulse response plotted in *Fig. 7.21* is shown in *Fig. 7.23*. The ETC is plotted on a dB magnitude scale versus time. The first arrival of the driver's impulse and the additional three reflection arrivals are

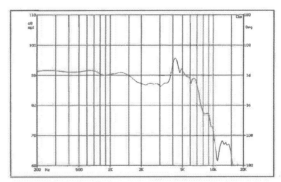

FIGURE 7.28: CLIO's frequency response of 7″ driver.

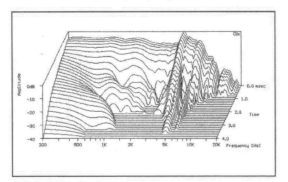

FIGURE 7.29: 7″ driver cumulative spectral decay.

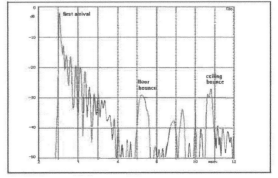

FIGURE 7.30: 7″ driver ETC with fiberglass.

indicated on the plot. Notice that the floor bounce is about 20dB down from the direct arrival. The ceiling bounce is down about 26dB.

The ETC is a useful *qualitative* measure of time of arrival. It is especially useful in the analysis of reflections in large acoustic spaces such as concert halls. Driver acoustic center is determined with much greater accuracy using the frequency-domain technique described in Section 6.8 and illustrated later in this chapter.

7.5.3 THE SYSTEM STEP RESPONSE

MLSSA derives an additional useful plot from the impulse response. This is the step response, i.e. the response to a step input of DC voltage. The step response of an ideal loudspeaker is shown in *Fig. 7.24*. Referring to the figure, a positive input-voltage step is applied to the voice coil at t = 1ms.

The driver cone immediately begins to move forward with a damped oscillatory motion, reaching a constant positive position beyond 15ms due to the now steady voice-coil current.

The sound pressure produced by this motion reaches the test microphone positioned 1m away roughly 2.9ms later. The pressure wave front produces a rapidly rising initial positive response followed by a damped oscillatory decay.

SPL actually undershoots the 0dB line and goes negative for a period of time, as it must, because the driver pressure response is high-pass. It cannot generate a DC SPL. The time average pressure response must be zero. This can only happen if the step response goes negative for some portion of the response interval.

The step response is a very useful *qualitative* measure of a loudspeaker system's time coherence. The system step response is the integral of its impulse response. MLSSA computes the step response by performing a numerical integration of the measured impulse response. *Figure 7.25* is a plot of the step response of a two-way MTM system employing a pair of 5.25″ midbass drivers with a centrally placed 28mm titanium dome tweeter.

The step is shown on an expanded time scale—the negative going portion beyond 5.2ms is not shown. Notice that there is an initial sharp positive peak in the step response followed by an undershoot and then a larger, but more slowly rising, positive peak. The first peak is produced by the tweeter, while the second peak is produced by the midbass driver pair. Although all drivers are connected with the same polarity, this system is not time coherent. The midbass arrival lags the tweeter by roughly 2.5ms.

This is a characteristic response for systems with higher-order passive crossover networks. With these crossovers, the woofer will lag the

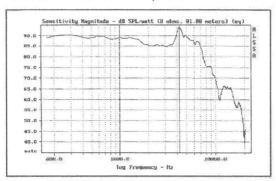

FIGURE 7.31a: 7″ driver response with cursor and marker at 0 and 6.2ms.

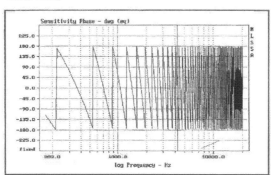

FIGURE 7.31b: Wrapped phase with DUT to microphone delay.

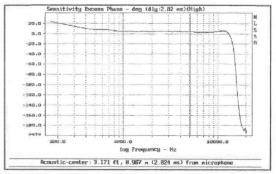

FIGURE 7.31d: Excess phase with DUT to microphone delay removed.

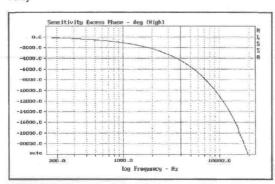

FIGURE 7.31c: Excess phase including driver to microphone delay.

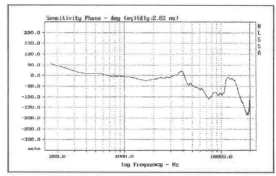

FIGURE 7.31e: 7″ driver phase response with delay removed.

tweeter even if their acoustic centers are in the same plane. These crossovers offer flat frequency response, but they are not linear phase. (Linear-phase crossovers have their own problems.)

7.5.4 THE CUMULATIVE SPECTRAL DECAY

The cumulative spectral decay (CSD) measures the frequency content of a system's decay response following an impulsive input. Ideally, a loudspeaker's impulse response should die away instantly. Real drivers, however, have inertia and stored energy which takes a finite time to dissipate. The CSD involves a series of frequency-domain calculations, but it is accessed from the time-domain menu in both MLSSA and CLIO.

Figure 7.26 shows how the CSD is computed. Successively shorter slices of the impulse response are used to compute a frequency response via an FFT. The first slice includes the entire impulse response out to a fixed end point which you can select by appropriate placement of a cursor. It is usually selected as that point in time just before the arrival of the first reflection. Succeeding slices are foreshortened toward this end point, including less and less of the early impulse response with each succeeding slice. The FFT of these slices yields the frequency content of later and later portions of the impulse response.

Figure 7.27 depicts the impulse response of a 7″ driver measured with CLIO. This plot should be compared against *Fig. 7.21*, which is the same driver as measured with MLSSA with one exception. Three layers of 6″ fiberglass were placed on the floor between the driver and microphone to reduce the effect of the floor bounce. Slowed by the fiberglass, the floor bounce moves out to 7ms and is reduced in amplitude by about 9dB (see the ETC discussion later).

The corresponding frequency response from 200Hz to 20kHz is shown in *Fig. 7.28*. This plot should be compared against the top curve in *Fig. 7.22*. The agreement between the two systems is quite good. The peak just above 4kHz represents a response resonance which should show up in the CSD.

The CSD of the 7″ driver is shown on the waterfall plot of *Fig. 7.29*. In this plot, frequency increases from left to right and time moves forward from the rear. The first 4ms of the CSD is shown with a total dynamic range of 40dB.

The rear curve labeled "0.0ms" analyzes the full impulse response, which is the same as the frequency response in *Fig. 7.28*. Well-defined ridges parallel to the time axis indicate strong decay modes. *Fig. 7.29* clearly shows a strong, long-lasting decay mode associated with the response peak just above 4kHz. Fortunately, when this driver is crossed over to a mating tweeter at 2kHz, this ridge is suppressed and does not appear in the full system response.

The ETC for this impulse response in *Fig. 7.27* is shown in *Fig. 7.30*. In comparison with *Fig. 7.23*, this plot shows that three layers of 6″ fiberglass have reduced the magnitude of the first floor bounce by 9dB. As you would expect, the

ceiling bounce remains unchanged. An FFT including the reduced amplitude floor bounce displays less ripple and can be smoothed to get a fairly good frequency-response plot.

Alternatively, a data window can be used to further suppress the effect of the floor bounce. Both CLIO and MLSSA provide a useful selection of data windows. Although this is not their primary purpose, these windows suppress the effect of the floor bounce by reducing energy in

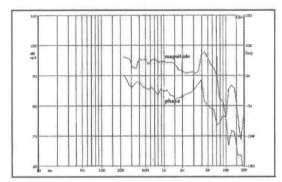

FIGURE 7.32: 7″ driver response with 1/6-octave sweep.

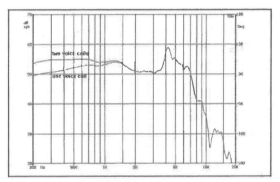

FIGURE 7.33: 7″ driver diffraction loss compensation.

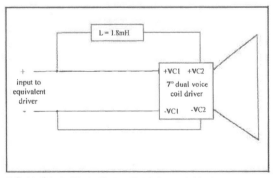

FIGURE 7.34: Equivalent single voice-coil driver.

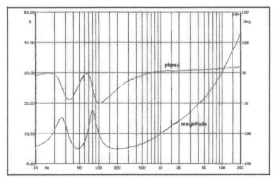

FIGURE 7.35: 7″ driver impedance in enclosure.

127

the tail of the impulse response. MLSSA provides an especially useful adaptive window.

Recall that a window decreases the effective resolution of an FFT by a smoothing action in the frequency domain. The Adaptive Window™ varies its length in the time domain as a function of frequency in order to maintain a constant percentage bandwidth smoothing. This window provides quasi-anechoic response at higher frequencies while letting more and more

of the room reflections in at lower frequencies. Compared to RTA analysis, this window produces a more accurate psychoacoustic measure of in-room response. An Adaptive Window example is presented in Section 7.8.

7.5.5 FREQUENCY AND PHASE RESPONSE

You have already seen some examples of frequency response computed from the impulse response, but a major advantage of PC-based acoustic data-acquisition and analysis systems like CLIO and MLSSA is that they also measure phase. One small problem was discussed in Chapter 6. Measuring loudspeaker phase accurately requires removal of the linear phase buildup due solely to the path length between the DUT and the measuring microphone.

CLIO will do this automatically when in the MLS mode by selecting the "delay" option in the MLS setup screen. CLIO does this using the procedure described in Section 6.8. Unfortunately, CLIO does not report the path

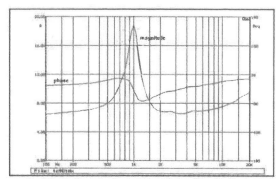

FIGURE 7.36: Titanium dome tweeter impedance.

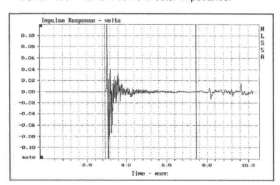

FIGURE 7.37: Tweeter impulse response at 1m.

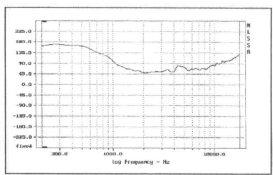

FIGURE 7.40: Tweeter phase response after delay removal.

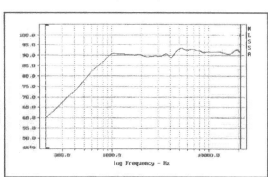

FIGURE 7.38: Tweeter 1m frequency response.

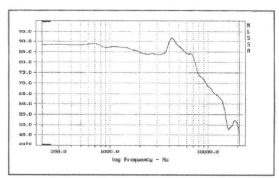

FIGURE 7.41: 7" woofer 1m far-field response.

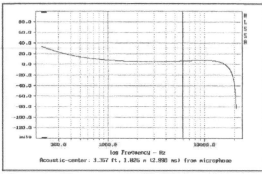

FIGURE 7.39: Tweeter excess phase after removing 2.998ms delay.

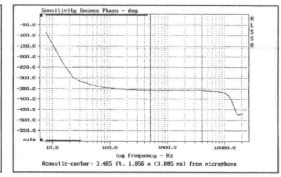

FIGURE 7.42: 7" woofer excess phase with delay removed.

LOUDSPEAKER
TESTING WITH
PC-BASED
ACOUSTIC DATA
ACQUISITION
SYSTEMS

delay it has removed nor the equivalent DUT to microphone spacing in the MLS mode.

With MLSSA, the path delay removal process is under operator control and all pertinent data is available to you. I'll take you through the steps for removing path delay and computing driver acoustic center with the 7″ driver previously discussed. Referring to *Fig. 7.21*, place the cursor at t = 0ms and the marker at 6.2ms. The analysis interval now includes the time-of-flight required to cross the driver to microphone gap.

The frequency response computed with this cursor and marker setting is shown in *Fig. 7.31a* from 200Hz–20kHz. It is essentially identical to the upper plot in *Fig. 7.22*. *Figure 7.31b* is a plot of the corresponding wrapped phase response. It shows a large number of +180° to –180° phase transitions. Executing the "View-Phase-Excess" command instructs MLSSA to compute the excess phase in the measured phase response. The result of this calculation is plotted in *Fig. 7.31c*. Excess phase exceeds 20,000° at 20kHz.

This driver will be used in a two-way system with a crossover frequency around 2kHz. We are there-fore interested in finding the delay which must be removed to make the driver minimum phase in a frequency range spanning at least one octave on either side of the crossover frequency. This is accomplished by first setting the marker and cursor on the excess phase graph to 1kHz and 4kHz, respectively (*Fig. 7.31e*). Then the "View-Phase-Delay" command is executed and followed with the "Calculate-Acoustic Center" command. The results of these two calculations are shown in *Fig. 7.31d*.

MLSSA has calculated a delay or time-of-flight of 2.824ms. With this time delay removed, the excess phase plot is now essentially zero in the 2–4kHz range (and actually all the way up to 12kHz). The time delay corresponds to a distance of 0.967m. Thus the acoustic driver center is 0.967m in front of the microphone. With the delay removed, the true driver phase response relative to the calculated acoustic center is shown in *Fig. 7.31e*.

CLIO will provide much the same information as MLSSA in its Sinusoidal menu under the Frequency response option. In this option CLIO will perform stepped sine wave response either with or without gating. Frequency steps of 1/3 to

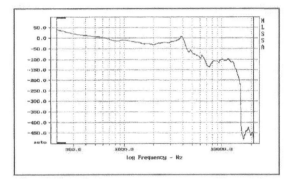

FIGURE 7.43: 7″ woofer unwrapped phase with delay removed.

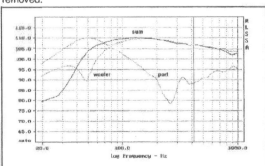

FIGURE 7.44: Near-field port and woofer (dotted), combined (solid).

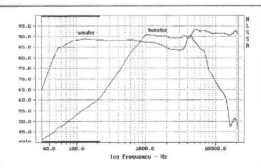

FIGURE 7.45: 7″ woofer and tweeter full-range responses.

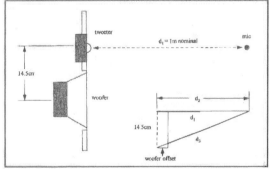

FIGURE 7.46: Geometry for determining driver acoustic phase center offsets.

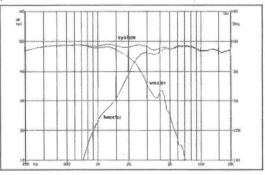

FIGURE 7.47: Two-way system and driver response.

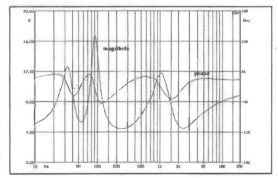

FIGURE 7.48: MTM ribbon system impedance.

1/48 octave are available. Lower and upper frequency limits within the range of 10Hz to 20kHz are also under operator control.

Gating can be used to reject data beyond a selected point in time. With the gating turned off, all room reflections will be included in the analysis. With the gating on you can, for example, reject all data after the point in time just before the floor bounce.

One remaining issue is when to turn on the gate. You would like the gate to be off during the time-of-flight so that environmental noise during that period does not contaminate the measurement. By selecting the "auto delay" option in the sinusoidal mode Setup screen, CLIO will automatically determine the signal time-of-flight and thus when to turn on the gate. You must select the frequency for this measurement.

Figure 7.32 shows the frequency and phase response of the 7″ driver measured with CLIO. The microphone placement was the same as that used to acquire the MLSSA data. Frequency was

stepped in 1/6 octave increments for this measurement. Frequency and phase response in this figure should be compared with the MLSSA results in *Figs. 7.31a* and *7.31e*. The agreement is quite good. The setup screen reported a gate turn-on delay of 2.83ms at 2kHz. This compares very well with MLSSA's delay of 2.824ms.

7.6 MEASUREMENTS FOR LOUDSPEAKER SYSTEM DESIGN

Modern crossover optimization software packages like those discussed briefly in Chapter 5 require very detailed information on driver impedance, frequency and phase response, and acoustic center location before the optimization process can begin. In my opinion, given the accuracy and sophistication of current crossover optimization software, generic manufacturer's data or model predictions are not good enough for design purposes. Design data must be specific to the acoustic environment in which the system of drivers will operate. For example, in earlier chapters you have

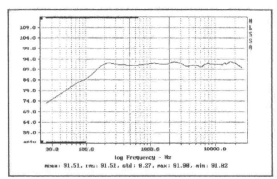

FIGURE 7.49: MTM ribbon system far-field quasi-anechoic response corrected to 1m.

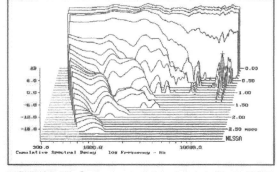

FIGURE 7.52: System cumulative spectral decay (CSD) response.

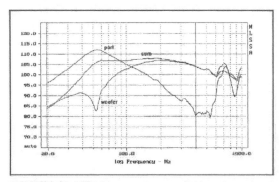

FIGURE 7.50: MTM ribbon system near-field woofer and port responses and their complex sum.

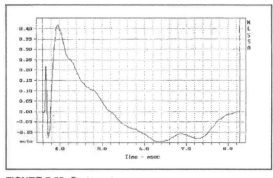

FIGURE 7.53: System step response.

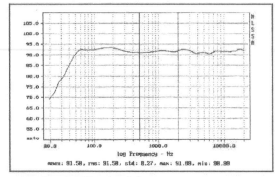

FIGURE 7.51: MTM ribbon full-range system response.

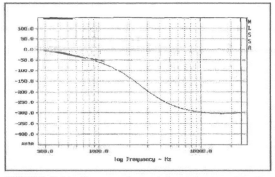

FIGURE 7.54: Excess phase delay referenced to the tweeter.

LOUDSPEAKER
TESTING WITH
PC-BASED
ACOUSTIC DATA
ACQUISITION
SYSTEMS

seen the very strong influence baffle geometry has on driver response. The data needed for the design process are:

- driver voice-coil impedance magnitude and phase
- driver on-axis frequency and phase response
- relative location of driver acoustic phase centers

The design process goes something like this: Based on measured T/S parameters and a desired bass loading (sealed, vented, bandpass, and so on), an overall enclosure volume is determined. An enclosure shape containing the required volume is then selected and a preliminary driver layout is made to permit construction of a prototype enclosure. Drivers are then mounted in the prototype enclosure to make the various measurements listed above.

I will illustrate this process with an example. The example involves a two-way system using a 7″ polykevlar cone woofer with a cast-aluminum frame and dual 8Ω voice coils. The tweeter is a 28mm inverted titanium dome unit. The woofer-driver T/S parameters were first measured with MLSSA.

The results indicated that this driver would work well in a vented QB3 alignment of 15 ltr net internal volume tuned to 49Hz.

A prototype enclosure was built with 0.75″ MDF and the drivers mounted flush on the front baffle centerline spaced 5.75″ (14.4cm) apart, center-to-center. A 2.5″ ID PVC pipe 5.75″ long tunes the enclosure to 49Hz.

This system will use the two woofer-driver voice coils in a somewhat unusual manner. One of the voice coils will be used for diffraction loss compensation. This coil will provide additional drive at low frequencies. An inductor in series with this voice coil will decrease its contribution to the output with increasing frequency. By proper selection of the series inductor, excellent compensation for the diffraction loss is possible.

Figure 7.33 shows the one-meter single voice-coil frequency response for the 7″ woofer in its enclosure taken with CLIO in the MLS mode. This response shows the characteristic low-frequency droop caused by diffraction loss. A sequence of frequency-response runs were next made with series inductors of 1.0, 1.2, 1.5, 1.8, and 2.2mH. Each

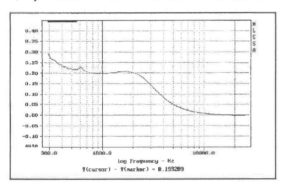

FIGURE 7.55: Excess group delay referenced to the tweeter.

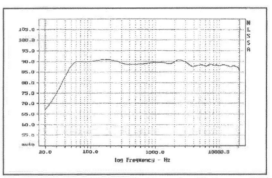

FIGURE 7.58: MTM response averaged over ±30° window

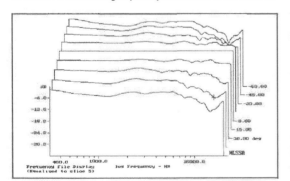

FIGURE 7.56: MTM horizontal polar response (–60° = 60° left).

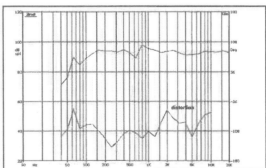

FIGURE 7.59: MTM second-harmonic distortion.

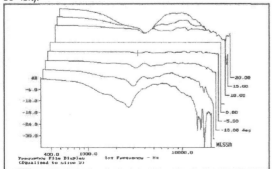

FIGURE 7.57: MTM vertical polar response (–° are down).

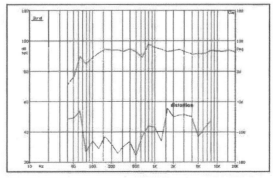

FIGURE 7.60: MTM third-harmonic distortion.

measurement takes only about 15s to complete. A 1.8mH coil produced the flattest response (also shown in *Fig. 7.33*).

At this point we have a diffraction-loss-compensated woofer driver which can be treated as a new driver without diffraction loss. The new driver is illustrated in *Fig. 7.34*. The impedance magnitude and phase of this composite driver were measured with CLIO using a sine-wave sweep with 1/24 octave steps in the 20Hz–20kHz range. The results of this measurement are plotted in *Fig. 7.35*. Impedance magnitude and phase of the 28mm tweeter were also measured with CLIO over the range of 200Hz to 20kHz in 1/12 octave steps. Tweeter impedance magnitude and phase are plotted in *Fig. 7.36*.

Now you must measure the on-axis frequency and phase response of each driver. For the tweeter, only a far-field quasi-anechoic measurement is needed. This measurement will extend down to 200Hz or so, which is sufficient for design purposes.

For the woofer driver, three measurements are required. First the far-field response is measured.

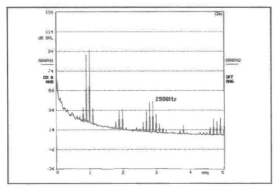

FIGURE 7.61: Dual woofer 900/1kHz intermods.

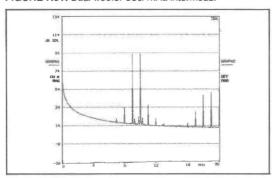

FIGURE 7.62: Tweeter intermodulation distortion.

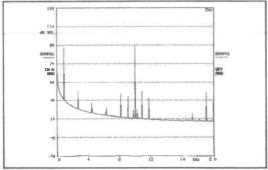

FIGURE 7.63: 900/10kHz intermodulation distortion.

Then woofer and port near-field responses are measured and combined to produce the complete low-frequency near-field response. Finally, the near-field and far-field responses are spliced at an appropriate frequency to get the full-range frequency response. The far-field data will also be used to calculate driver-acoustic center positions.

All driver-response measurements for this example were made with MLSSA. The microphone was placed on the tweeter axis at 1m and was not moved for subsequent measurements. A bubble level was used to ensure that the baffle was truly vertical.

The 1m tweeter impulse response is shown in *Fig. 7.37*. The marker and cursor bracket a 4.5ms quasi-anechoic time interval in the impulse response. The tweeter's quasi-anechoic frequency response is shown in *Fig. 7.38*. It is valid down to 220Hz. Sensitivity between 1–3kHz averages 90dB SPL/2.83V/1m. There is a 2dB rise in tweeter response between 4–5kHz, but this will disappear in the final design.

Figure 7.39 shows tweeter excess phase with a 2.998ms delay removed from the impulse response. MLSSA calculates the tweeter's acoustic phase center at 1.026m from the microphone. At this distance the tweeter is minimum phase over a frequency range of 1–12kHz. Tweeter phase response with the excess phase removed is plotted in *Fig. 7.40*.

Relative to the calculated acoustic center, the tweeter phase never goes negative. It actually bottoms out in the 2–3kHz range and then begins to rise. This is caused by the tweeter's ultrasonic resonance at 25kHz, where response peaks about 10dB. This peak is not shown in *Fig. 7.38*, which extends out only to 20kHz.

Woofer-driver quasi-anechoic frequency response is shown in *Fig. 7.41*. Like the tweeter response, it is valid down to 220Hz. This is the same frequency response as that shown in the upper curve of *Fig. 7.31a*, but here it has been smoothed by 1/6 octave. Woofer-driver excess phase after delay removal is plotted in *Fig. 7.42*. MLSSA removed a 3.085ms delay to reduce excess phase to zero in the 2–4kHz range. (The excess phase in this range actually reads −360°, but this is the same as zero degrees.)

Unwrapped woofer-driver phase response with all excess phase removed is shown in *Fig. 7.43*. Notice a rapid buildup in the woofer-driver phase above 12kHz. This is caused by the sharp drop in the frequency response and a build up of excess phase above this same frequency.

Near-field woofer and port responses are plotted in *Fig. 7.44* along with their complex sum, which represents the total low-frequency near-field response. Recall in Chapters 4 and 5 examples of near-field port and woofer responses, which could not be added together without phase information. MLSSA and CLIO provide auxiliary math operations to perform the addition. The operation is most direct in MLSSA, which performs a weighted average of the two responses.

As part of the command sequence, MLSSA asks for the weighting to be applied to each response in the weighted sum. The driver and port diameters

LOUDSPEAKER
TESTING WITH
PC-BASED
ACOUSTIC DATA
ACQUISITION
SYSTEMS

are the weighting factors. You can also add a negative delay to the port phase response if the port mouth is very far from the woofer. This would be the case, for example, with a port opening located on the rear of the enclosure.

The full-range woofer-driver and tweeter responses are plotted in *Fig. 7.45*. The woofer response is obtained by splicing the near-field and far-field responses at 220Hz. With CLIO this is done in the MLS mode with the Merge command under the Math submenu. In MLSSA you invoke the spLice command in the frequency domain.

Both systems match the low-frequency response to the far-field response. Both magnitude and phase are matched so that they are continuous across the splice. *Figure 7.45* shows that the woofer and tweeter responses overlap between 1–4kHz, suggesting a crossover frequency in the 2–2.5kHz range. One additional point: Notice that the near-field woofer response shows a deep notch just below 50Hz, indicating the vented box frequency.

Completing the design data requires determining the woofer acoustic-phase-center offset relative to the tweeter. Recall that both drivers are flush mounted to the baffle, which was then set to true vertical with a bubble level. After taking the tweeter measurement, you could reposition the microphone in front of the woofer at exactly 1m. In practice I have found this rather difficult to do reliably. It is much simpler to keep the microphone location fixed on the tweeter axis at a nominal 1m distance and use a little trigonometry to get the correct offset. The exact driver-to-microphone spacing is not critical. It is only important that the microphone position not change.

The tweeter-woofer-microphone geometry is shown in *Fig. 7.46*. The woofer is being measured slightly off-axis, but as long as this angle is small, there is little error in the resulting frequency response. In this example the angle is 8°. The distances measured with MLSSA are:

$$d_1 = 102.6cm \text{ and } d_3 = 105.6cm$$

d_2 is the distance that would be measured if the two driver axes were in line.

$$d_2 = \sqrt{{d_3}^2 - (14.5)^2} = \sqrt{(105.6)^2 - (14.5)^2} = 104.6cm$$

Now the interdriver phase-center offset, Δd, is simply:

$$\Delta d = d_3 - d_2 = 104.6 - 102.6 = 2cm$$

Computing Δd essentially involves differencing the two measurements, d_1 and d_3. This approach is quite accurate since any uncompensated delay or phase errors are common to both measurements and cancel in the differencing operation.

In making the above calculation, I have assumed that the driver acoustic-phase-center lies on the driver's axis. The phase centers will be aligned when listening to the system below the tweeter's axis by an angle Φ, where

$$\Phi = \arctan(2/14.5) = 7.9^0$$

This angle can be realized by tilting the enclosure back by the same angle. The crossover design software can be set up to optimize response along this tilted axis.

All of the information needed to design a crossover for this two-way system is available now. A fourth-order *acoustic* Linkwitz-Riley crossover was designed for this system using the XOPT crossover optimization software. The resulting woofer and tweeter crossovers required to produce a fourth-order acoustic response are second- and third-order electrical filters, respectively. A mockup of the crossover was constructed and frequency-response tests conducted.

The far-field on-axis system response and the response of the individual drivers are shown in *Fig. 7.47*. Crossover occurs at 2275Hz. The system response is flat within ±1.5dB from 200Hz to 20kHz. Notice that the woofer response peak has been reduced by about 12dB.

7.7 MEASUREMENTS FOR LOUDSPEAKER SYSTEM PERFORMANCE ANALYSIS AND EVALUATION

The measurements needed for loudspeaker system analysis and evaluation are quite different from those required for design purposes. Within their limitations, measurements for analysis and evaluation should help to determine how a system will sound when placed in a typical listening environment. The measurements which in my experience relate most directly to this goal are:

- system impedance magnitude and phase
- on-axis frequency response
- system sensitivity

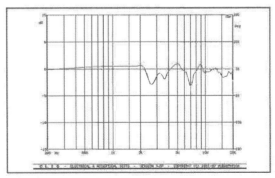

FIGURE 7.64: Effect of grille on frequency response.

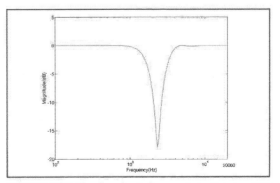

FIGURE 7.65: Fourth-order L-R crossover response with tweeter delay.

133

- cumulative spectral decay
- step response
- excess group delay
- horizontal and vertical polar response
- power response
- harmonic and intermodulation distortion

The following example will illustrate the application of these measurements to the performance analysis and evaluation of a loudspeaker system,

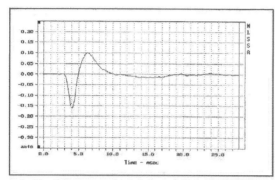

FIGURE 7.66: Impulse response at the end of the transmission line.

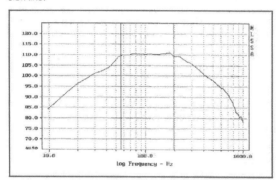

FIGURE 7.67: Frequency response at end of transmission line.

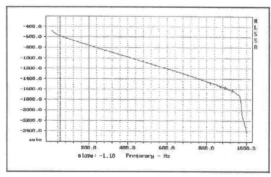

FIGURE 7.68: Transmission line excess phase.

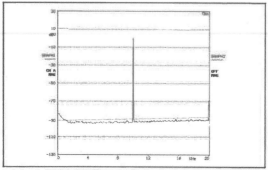

FIGURE 7.69: 10kHz sine wave FFT with no window.

TABLE 7.6
CHARACTERISTICS OF WINDOWED SPECTRA

window	bin levels(dBV) (10.0 & 10.1kHz)	–90dB width (kHz)
rectangular	–3.8	——
Hanning	–7.4	3.5
Blackman	–8.6	2.5

which consists of an MTM array similar to the one in Fig. 5.24. The woofers are 5.25″ units with cast aluminum frames, a polykevlar sandwich cone, a neoprene half-roll surround, and a vented pole piece. The tweeter is a ribbon type, 6cm long and 7mm wide, with the longer dimension oriented vertically. The system uses a vented alignment with a fourth-order acoustic Linkwitz-Riley crossover.

First, measure system impedance with CLIO (*Fig. 7.48*). The impedance minimum of 5.2Ω at 54Hz indicates the resonant frequency of the vented box. An overall minimum impedance of 4.4Ω occurs at 240Hz. There is a final minimum at 2350Hz, which has a value of 4.9Ω.

The capacitive impedance of the tweeter crossover network combines with the inductive impedance of the woofer pair crossover to form a parallel resonance condition just above 1kHz. There is also a small glitch in the impedance curve just below 1kHz, which is probably caused by a standing-wave mode in the enclosure. Impedance phase ranges from $-40°$ to $+29°$ over the full audio range. This should be a relatively easy load for most amplifiers.

Figure 7.49 shows the system's far-field, quasi-anechoic on-axis frequency response. The microphone was placed on the tweeter axis at a distance of 48″ for this measurement to assure proper inter-driver integration. The far-field response is valid down to about 220Hz. The measurement has been normalized to 1m to get system sensitivity. System sensitivity in the two octaves around 1kHz (500Hz to 2kHz) averages 91.5dB/2.83V/1m.

The system is tuned to 54Hz with two 1.5″ ID ports. The near-field responses of one port and one woofer are shown in *Fig. 7.50*. The dip in woofer response just above 50Hz also indicates the box tuning frequency. The port output is near maximum at this point.

The port curve also shows a peak at 720Hz, which is the result of an "organ-pipe" resonance in the port tube. This resonance causes a dip in the overall near-field response, but because the port exit is on the rear of the enclosure, it does not show up in the far-field response of *Fig. 7.49*. If the speaker is positioned with its back close to a reflecting surface, the organ-pipe resonance could become audible.

The port and woofer near-field responses are summed by the MLSSA system, giving proper weighting to the difference in areas of the combined woofers and ports, to obtain the complete near-field low-frequency system response. This response is also shown in *Fig. 7.50*. The near-field response is then spliced to the quasi-anechoic response at 220Hz to get the complete system

LOUDSPEAKER
TESTING WITH
PC-BASED
ACOUSTIC DATA
ACQUISITION
SYSTEMS

response (*Fig. 7.51*). On-axis response is flat within ±1.6dB from 60Hz to 20kHz.

The system cumulative spectral decay (CSD) response is shown in the waterfall plot of *Fig. 7.52*. The first three milliseconds of the CSD are plotted with a total dynamic range of 32dB. The tweeter response is essentially gone in 0.3ms. Beyond 0.3ms there is some small amount of residual hash in the 12–14kHz range down about 18dB which persists out to 1.4ms, but it has no special character which might indicate a delayed tweeter resonance. The tweeter decay response is excellent.

Decay response below 3kHz is controlled by the woofer and its crossover network. Here again decay response is quite good. There are no distinct unimodal ridges that would indicate the presence of strong system resonances.

Figure 7.53 is a plot of system step response. The initial positive spike indicates the tweeter arrival. It is followed by the woofer pair arrival, peaking about 0.3ms later. The drivers are connected with positive polarity, but the system is not time coherent.

To get a better view of this behavior, first compute system excess phase and reference it to the tweeter location. Then an excess group delay plot will show the time separation between the woofer pair and the tweeter.

System excess phase is plotted in *Fig. 7.54*. The cursor and marker were set at 10 and 20kHz, respectively (the ribbon tweeter's response extends out to 40kHz), and the correct delay removed to reference the excess phase calculation to the tweeter's acoustic center. The resulting excess group delay is shown in *Fig. 7.55*. In a time-coherent system this plot would be a flat line. Above 10kHz excess group delay is essentially zero, as it should be since it is referenced to the tweeter in this frequency range.

The curve rises below 10kHz to a plateau starting just below 2kHz. The difference in excess group delay between 15kHz and 1kHz points is 0.199ms, or 199μs. This is the woofer pair time delay relative to the tweeter. This time offset of the drivers is a direct consequence of the fourth-order acoustic Linkwitz-Riley crossover used in this system. Excess group delay is a very accurate indicator of driver time offset. The perceptual effect of this lack of time coherence is a hotly debated topic with no clear conclusions.

MLSSA has a very useful feature for visualizing loudspeaker polar response. Frequency responses taken at successive off-axis angles can be plotted in a waterfall view. This presentation produces a polar-response surface, which reveals polar-response anomalies at a glance.

For good stereo imaging and proper spectral balance from side wall reflections, the horizontal off-axis curves should be smooth replicas of the on-axis response with an allowable exception for the natural rolloff of the tweeter at higher frequencies and wider angles. (Remember: Our ear-brain combination rejects higher frequency sidewall reflections.)

Figure 7.56 is a waterfall plot of horizontal polar response taken in 15° increments from 60° left to 60° right when facing the speaker. The micro-

phone is placed at tweeter height for this series of curves. All off-axis plots are referenced to the on-axis response, which appears as a straight line at 0.00°. Off-axis plots show the change in response relative to the on-axis response.

There are no polar-response anomalies in the crossover region, indicating a smooth transition from the woofer pair to the tweeter. You can clearly see the expected rolloff of tweeter response at higher frequencies and larger off-axis angles. Tweeter response is down 10.1dB at 15kHz and ±45°. The corresponding figures at ±30° and ±15° are 4.7 and 1.7dB, respectively.

This is as good as many dome tweeters and better than most that I have tested. The tweeter's ribbon is recessed slightly within the slot formed by its magnetic gap. Tweeter output diffracts around this slot producing the uptick in response seen above 15kHz at horizontal off-axis angles of ±45° and ±60°.

Figure 7.57 is the waterfall plot of vertical polar response. Responses are shown in 5° increments from 20° below (−20°) the tweeter axis to 20° above it. Response at ±5° is within 1dB of the on-axis response. Worst-case response dips at ±10° and ±15°

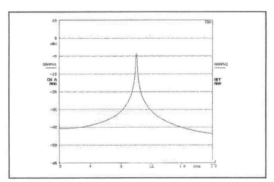
FIGURE 7.70: 10,050Hz FFT with no window.

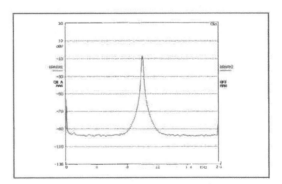
FIGURE 7.71: 10,050Hz FFT with Hanning window.

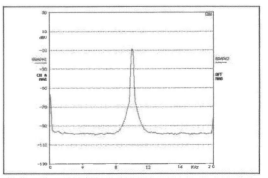
FIGURE 7.72: 10,050Hz FFT with Blackman window.

135

are 4.1 and 8.8dB, respectively. Both of these dips are in the crossover region as expected, due to the vertical orientation of the system drivers.

The tweeter ribbon is 6cm long in the vertical direction, which is equal to one wavelength at 5.7kHz. Above this frequency the tweeter will become more directional in the vertical. This explains the rapid drop in high-frequency response above 5kHz at ±15° and beyond.

The power response is determined by averaging the response over a 60° horizontal angle (±30°) in the forward direction. Both CLIO and MLSSA provide auxiliary math functions to perform this averaging. The power response (*Fig. 7.58*) is very smooth, sloping gently downward by about 2dB above 3kHz. This is excellent horizontal polar response and suggests good direct field coverage in the primary listening area with little if any timbral change. Image stability should be very good.

CLIO was used to measure harmonic and intermodulation distortion. Harmonic-distortion tests were run at an average SPL of 90dB at 1m. *Figures 7.59* and *7.60* show second- and third-harmonic distortion levels in dB SPL versus frequency in 1/3 octave steps. System frequency response is also plotted on these figures.

The second- and third-harmonic distortion figures at 50Hz are 35 and 27dB below the full system SPL, respectively. This corresponds to second- and third-harmonic distortions of 1.8% and 4.6%. The corresponding numbers at 100Hz are 0.6% and 0.2%. All woofer distortions are well below 1% above 100Hz. Considering the size of these woofers, this is excellent performance.

Harmonic distortion does rise above 1% briefly at 2kHz. This distortion comes from the tweeter, which could benefit from a somewhat higher crossover frequency. But this would restrict vertical coverage in the crossover region. The world is full of compromises, and there is no free lunch!

Intermodulation distortion is easily measured with CLIO's FFT mode analysis and its two-tone test signal. In this type of test two nearby frequencies are input to the speaker. Nonlinear speaker response then creates output frequencies, which are not harmonically related to the input. These frequencies are much more audible and annoying than harmonic distortion.

Let the symbols f_1 and f_2 represent the two frequencies used in the test. Then a second-order nonlinearity will produce intermods at frequencies

of $f_1 \pm f_2$. A third-order nonlinearity generates intermods at $2f_1 \pm f_2$ and $f_1 \pm 2f_2$.

Woofer intermods were examined first by inputting 900Hz and 1kHz signals to the system at equal levels. These frequencies are far enough below the crossover that little if any energy will leak into the tweeter. Total SPL with the two signals was adjusted to 90dB at 1m. The output spectrum from this test is shown in *Fig. 7.61*. The two tallest lines represent the input signals. The following lists the primary intermodulation distortion products and how they are produced:

$$1900\text{Hz} = 900 + 1000 \quad \text{(second-order)}$$
$$800\text{Hz} = 2 \times 900 - 1000 \quad \text{(third-order)}$$
$$1100\text{Hz} = 2 \times 1000 - 900 \quad \text{(third-order)}$$
$$2800\text{Hz} = 2 \times 900 + 1000 \quad \text{(third-order)}$$
$$2900\text{Hz} = 2 \times 1000 + 900 \quad \text{(third-order)}$$

Other lines on the plot are due to harmonic distortion, which we have already discussed. The majority of intermods are third-order, but the largest distortion product at 2900Hz is 54dB below the main output, which is equal to 0.2%. This is better than some solid-state amplifiers and most tube amps!

Tweeter intermods were measured with a 10 and 11kHz input pair also adjusted to produce 90dB SPL (*Fig. 7.62*). Intermods are at 9, 11, 12, and 19kHz. The worst case tweeter intermod at 19kHz is down 46dB at 0.5%. This is very good performance.

The last intermod test examines cross-intermodulation between the woofers and tweeter using frequencies of 900Hz and 10kHz. (A 1kHz signal would produce intermods that fall on harmonic-distortion lines and confuse the results.) This spectrum is shown in *Fig. 7.63*. Intermods can be seen at 8.2, 9.1, 10.9, and 11.8kHz. The highest level at 11.8kHz is 51.5dB down at 0.3%.

Ideally, there is no mechanism by which 900Hz and 10kHz signals can mix. But in practice cross coupling in the crossover networks and the common ground wire introduce low frequencies into the tweeter and high frequencies into the woofers. This makes the case for biwiring.

Cross-intermodulation could also occur with open-backed tweeters or tweeters with vented pole pieces that are not isolated from the woofer backwave. Low-frequency pressurization of the enclosure by the woofer would then modulate tweeter cone excursion. Nonlinear mixing can occur in

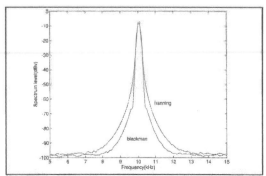

FIGURE 7.73: Comparison of Hanning and Blackman windows.

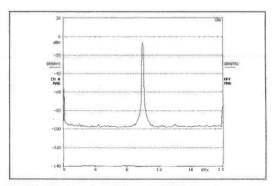

FIGURE 7.74: Hanning windowed 10kHz sine wave.

LOUDSPEAKER
TESTING WITH
PC-BASED
ACOUSTIC DATA
ACQUISITION
SYSTEMS

the air at extremely high SPLs, but this is not the case here.

All of the above tests were conducted with the grille off. *Figure 7.64* shows the system response with the grille on, but referenced to the response with the grille off; that is, it plots the *difference* in response under the two conditions. Below 2kHz the grille has no significant effect. Above 2kHz, however, the grille causes rather ragged response deviations of 4dB peak-to-peak. The grille for this system is only for cosmetics.

7.7.1 A COMMENT ON ALL-PASS CROSSOVERS

I must digress from the main topic of this chapter for a moment. Both this chapter and Chapter 6 illustrate that allpass crossovers produce a time off-set between the woofer and tweeter. All-pass crossovers sum to flat frequency response when all driver acoustic phase centers are aligned. The time offset is caused by the crossover network itself and is not due to any driver offsets. This is the price paid for the flat frequency response and higher attenuation rates of higher-order all-pass crossovers, which are not time coherent.

The time offset is clearly shown in the excess group-delay plots. Recall that excess group delay is the derivative of excess phase. Wherever excess group delay is flat, the excess phase curve is linear. The flat intervals are frequency regions where the system can be made minimum phase with the proper time delay.

Looking at the excess group delay plot of *Fig. 7.55*, you might think that delaying the tweeter by 199μs would bring the entire system into time coincidence and produce an ideal loudspeaker. This could be accomplished by moving the tweeter back along its axis by 6.8cm. Unfortunately, time coincidence can only be attained at the cost of frequency response.

Figure 7.65 shows the summed response of fourth-order woofer and tweeter crossover responses with the tweeter output delayed by 199μs. Response dips about 18dB in the crossover region. Delaying the tweeter does not make the system minimum phase, and the effect on frequency response is unacceptable.

7.8 ADDITIONAL INTERESTING EXAMPLES

In this section I will discuss three additional topics that illustrate the broad utility of PC-based acoustic and electrical data-acquisition systems. First, I'll further analyze the transmission-line example first encountered in Chapter 4, then take another look at windows and leakage with CLIO's help, and, finally, examine a very special window and its use in room/loudspeaker analysis and equalization.

7.8.1 ANOTHER LOOK AT THE TRANSMISSION LINE

In this section I will use frequency-domain concepts to compute the sound velocity in a stuffed transmission line. *Figure 7.66* is a plot of the impulse response measured at the output end of the transmission line. For this measurement the microphone was placed in the plane of the transmission-line exit port. It takes roughly 3ms for the

leading edge of the impulse to arrive at the line's exit. The first arrival is negative going as you would expect, since it comes off the rear of the cone.

The frequency response corresponding to this impulse response (*Fig. 7.67*) is quite flat from 50Hz–200Hz and falls off at a rate of roughly 9dB/octave on either side of this range. Now look at *Fig. 7.68*, which shows the transmission-line's excess phase plotted on a linear frequency scale. The excess phase slope is almost a straight line from 40Hz–800Hz. Over this range the transmission line is essentially a pure time delay.

MLSSA has a broad selection of auxiliary statistics functions. One function will estimate the slope of the excess phase plot. I placed a cursor and marker on the excess phase plot of *Fig. 7.68* to bracket the region of flat frequency response shown in *Fig. 7.67*. The slope estimate for this range is $-1.18°/Hz$. This slope must be converted to rad/rad/sec, which has the units of time. The conversion goes like this:

$$TL \text{ delay} = 1.18 \left[\frac{\deg}{Hz} \times \frac{\pi F}{180°} \times \frac{1 Hz}{2\pi \text{rad}/s} \right] = \frac{1.18}{360} s = 3.28 ms$$

This time delay corresponds to an in air-path length, TL_{length}, of

$$TL_{Length} = (3.28 \times 10^{-3} \text{ sec}) \times (34,400 cm/s) = 112.8 cm = 44.4''$$

The *physical* length of the line measured along its centerline is $34''$. Thus the line stuffing material slows the speed of sound to 34/44.4, or 76% of its value in air. This is very close to the theoretical minimum air speed of 71% for an isothermal

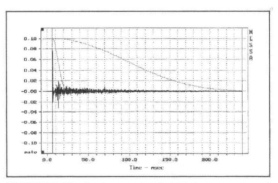

FIGURE 7.75: Room/speaker impulse response and Adaptive Window.

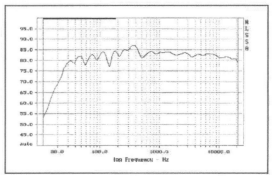

FIGURE 7.76: Room/speaker response with Adaptive Window.

medium.

The concept of excess phase has allowed us to compute a very accurate value for the speed of sound in a filled transmission line. This could not have been done as accurately in the time domain. Deciding when the pulse has arrived by looking at *Fig. 7.66* is somewhat subjective. The impulse response gives you only a rough measure of the delay. The step response and ETC are no better, but the frequency-domain result is straightforward, accurate, and unequivocal.

7.8.2 ANOTHER LOOK AT SPECTRAL LEAKAGE AND WINDOWS

CLIO's FFT mode gives a very graphic look at the effect of windows on spectral leakage. The sampling parameters for all the results to be shown are:

$$F_S = 51.2kHz \qquad L = 512 \text{ samples} \qquad f_0 = 100Hz$$

Figure 7.69 is the spectrum of an unwindowed 10kHz sine wave at 0dBV. Since 10kHz is an integer multiple of f_0, there is no leakage. (The non-zero width at the base of the spectral line is an artifact of the plot.)

The spectrum of a 10,050Hz sine wave is plotted in *Fig. 7.70*. This frequency falls in the middle of the 10,000Hz and 10,100Hz bins. The spectral amplitude in each of these bins is equal at -3.8dBV. The power in the 10,050Hz sine wave has been split equally between the two nearest bins. Notice that the scale of this plot is 10dB/div versus 20dB/div for *Fig. 7.69*. The spectrum decays slowly to either side of the two maxima, with a maximum attenuation at the ends of the analysis range of a little over 45dB.

Figures 7.71 and *7.72* show the reduction is spectral leakage attained with Hanning and Blackman windows. *Table 7.6* summarizes the pertinent characteristics of the three spectra. The unwindowed spectrum gives equal weight to all points in the sample. This is equivalent to using a rectangular window with a constant height of one. Notice that the spectral levels at 10kHz and 10.1kHz are lower in the windowed spectra than the unwindowed spectrum. As noted in Chapter 6, the window reduces the total power in the sample.

Figure 7.73 plots the Hanning and Blackman windowed spectra on the same graph. Here the Blackman window has somewhat narrower

skirts than the Hanning window, but more power is lost with the Blackman window. The Hanning window is especially useful for windowing impulse response data. The Blackman window is optimized for power spectrum estimation.

There is an additional price to pay for leakage reduction. The spectrum of windowed time-domain data has poorer resolution than an unwindowed spectrum. *Figure 7.74* is the spectrum of a 10kHz sine wave that has been Hanning windowed. Instead of a single line at 10kHz, the spectrum is now spread out over several neighboring bins.

The act of windowing data in the time domain is equivalent to passing a smoothing filter over the unwindowed spectrum, much like the 1/3 octave smoothing of frequency-response plots. There is a difference, however. The smoothing filter produced by time-domain windowing has a constant bandwidth, or, equivalently, a constant rise time. This contrasts with the 1/3 octave filter or 1/n octave filters in general that have constant *percentage* bandwidth. In the next section I will discuss an Adaptive Window™, which is especially useful in generating room/speaker response characteristics.

7.8.3 COMBINED ROOM/LOUDSPEAKER RESPONSE

Efforts so far have been aimed at removing environmental effects from measurements. However, loudspeakers must ultimately operate in real rooms, so measurements should consider meaningful ways to determine in-room response with an eye towards room/loudspeaker equalization. I have previously discussed the use of 1/3 octave RTA measurements to quantify the in-room response of a loudspeaker. You can also get the in-room loudspeaker response by including more and more of the loudspeaker impulse response tail in the analysis. For example, you could use 100–200ms of the impulse response to compute frequency response. Then, 1/3 octave smoothing of this spectrum should give something similar to RTA analysis.

MLSSA takes a different approach to in-room speaker response measurement, using an Adaptive Window™ on the impulse response. The Adaptive Window is said to produce in-

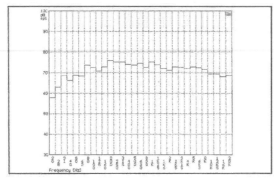

FIGURE 7.77: 1/3-octave RTA.

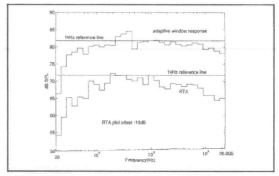

FIGURE 7.78: Comparison of RTA and Adaptive Window room/speaker responses.

LOUDSPEAKER
TESTING WITH
PC-BASED
ACOUSTIC DATA
ACQUISITION
SYSTEMS

room response measurements that more closely reflect listeners' perceptions of spectral balance when compared to RTA measurements. Listeners tend to ignore late room reflections in the treble region while giving them increasingly more weight with decreasing frequency. The Adaptive Window imitates this behavior by adaptively varying its length as a function of frequency. This window provides near quasi-anechoic response at higher frequencies while letting more and more of the room reflections in at lower frequencies.

The in-room response of the two-way loudspeaker described in Section 7.6 was measured with both CLIO and MLSSA. The CLIO measurement was made with pink noise in the RTA mode. With MLSSA the Adaptive Window was used. The loudspeaker was set up about 3' in front of the rear wall and far from all side walls in a large room. The microphone was placed 8' in front of the loudspeaker at tweeter height.

Figure 7.75 shows the first 240ms of the impulse response and the two extremes of the Adaptive Window. The Adaptive Window algorithm varies the length of the data window applied to the impulse response smoothly and continuously between the two extremes shown in the figure. The short window is normally set to include the first 10ms or so of the impulse response. This covers the quasi-anechoic range and perhaps one or two of the earliest room reflections. The long window determines the ultimate low-frequency limit of the analysis.

The Adaptively Windowed frequency response includes only the earliest reflections at high frequencies. As you go down in frequency, it includes more and more room reflections, finally including almost all reflections at the lowest frequency. The maximum window length is set to one-half the FFT length. It is advisable to keep the longer window under 200ms in length. Reflections arriving beyond this point are perceived as distinct echoes.

Figure 7.76 shows the 1/3 octave smoothed Adaptive Window response. This plot should be compared to the full-range response in *Fig. 7.51*. CLIO's RTA response is shown in *Fig. 7.77*. To ease comparison of the two results, I also sampled the 1/3 octave smoothed MLSSA curve at intervals corresponding to the center frequencies of the 1/3 octave bands starting at 20Hz. Both results are plotted on the same graph for easy comparison (*Fig. 7.78*). A reference line corresponding to the response level at 1kHz is added to both plots.

Notice that the RTA response falls off more rapidly above 1kHz relative to the Adaptive Window response. This is caused by the directional properties of the measurement microphone. Even omnidirectional microphones will miss some of the high-frequency reverberant sound arriving far off-axis. As a result, the RTA values at 1kHz and below are falsely elevated in level relative to the higher frequencies. This may explain why room/speaker equalization using RTA measurements often sound brighter

than they should.

The Adaptive Window retains phase information in the response measurement as does simple 1/3 octave smoothing of MLSSA or CLIO MLS response curves. This may become important as digital room equalizers become more popular. MLSSA will compute the inverse equalization curve for you. This curve can then be programmed into an equalizer.

7.9 CONCLUSION

I hope this chapter has convinced you of the power and convenience of PC-based acoustical and electrical data-acquisition and analysis systems. What is not readily apparent when reading this text is the speed with which measurements can be made. MLS impulse response and impedance measurements can be made in seconds. Sinusoidal frequency sweeps may take one or two minutes, but this still beats the point-by-point approach with analog test gear. Almost everyone reading this book has a PC. For less than the cost of that PC, the technician or serious hobbyist can equip him/herself with a PC-based data-acquisition system.

REFERENCES
1. D. Rife, "MLSSA Speaker Parameter Option," Reference Manual, Version 4.0A, DRA Laboratories, 1991–1996.
2. D. Campbell, "MLSSA vs. Sine Testing: A Tutorial on Methodology," *SB* 1/92, January 1992.
3. D. Gabor, "Theory of Communications," Proceedings of the IRE, Vol. 93, 1946.
4. R.C. Heyser, "Determination of Loudspeaker Arrival Time, Parts I, II & III," *JAES*, Vol. 19, Numbers 9, 10, and 11, October, November, and December, 1971.
5. J. Vanderkooy and S. Lipshitz, "Uses and Abuses of the Energy-Time Curve," *JAES*, Vol. 38, No. 11, pp. 819–836, November 1990.
6. D. Rife, "MLSSA, Maximum-Length Sequence System Analyzer," Reference Manual, Version 10.0A, Section 3.11, copyright DRA Laboratories, 1987–96.

ABOUT THE AUTHOR

Joseph D'Appolito is an internationally recognized authority in audio and acoustics, specializing in loudspeaker system design. He has consulted with organizations world wide, including in the United States, Canada, Europe, and the Far East. He originated the popular MTM loudspeaker geometry, commonly known as the "D'Appolito Configuration," which is now used by dozens of manufacturers throughout Europe and North America.

As a member of the Audio Engineering Society, Dr. D'Appolito has published conference and journal papers demonstrating his in-depth understanding of loudspeaker system and crossover design. Using computer-based tools and specialized test equipment, he has designed many highly successful loudspeaker systems, including the ARIA 5 Point Source for Focal (France), which was selected loudspeaker of the year for 1991 by *Hifi Video* magazine (Paris).

As a Contributing Editor to *Speaker Builder*, Dr. D'Appolito has contributed widely to the popular literature on loudspeaker design and testing over the last two decades. He designed Mitey Mike and its successor, Mitey Mike II, which are low-cost, precision microphones for loudspeaker testing. He also developed TopBox loudspeaker design software. He is Contributing Technical Editor for three popular books on home and auto loudspeaker construction published by Master Publishing for Radio Shack.

Dr. D'Appolito holds BEE, SMEE, EE, and Ph.D. degrees in electrical engineering from RPI, MIT, and the University of Massachusetts. In addition to his specialization in acoustics, he has authored over 30 journal and conference papers on applications of modern estimation and control theory, statistical signal processing and stochastic systems identification and modeling to a broad range of navigation, guidance, communications, and signal processing problems. He now runs his own consulting firm specializing in audio and acoustics.

INDEX